T5-BYZ-054

FOR REFERENCE
NOT TO BE TAKEN FROM THE ROOM
CAT. NO. 23 012
PRINTED IN U.S.A.

Taxonomic Terminology of the Higher Plants

TAXONOMIC TERMINOLOGY of the HIGHER PLANTS

By H. I. Featherly

Professor of Botany

Oklahoma A. & M. College

(Facsimile of the Edition of 1959)

Hafner Press
A Division of Macmillan Publishing Co., Inc.
New York
Collier Macmillan Publishers
London

Reprinted by Arrangement
1965
Second Printing 1973

HAFNER PRESS
A Division of Macmillan Publishing Co., Inc.
866 Third Avenue, New York, N. Y. 10022
Collier-Macmillan Canada, Ltd., Toronto, Ontario

Library of Congress Catalog Card Number: 53-12176
ISBN: 02-844590-2

Printed in U.S.A. by
NOBLE OFFSET PRINTERS, INC.
New York, N.Y. 10003

To the memory of my Mother,
whose life was a constant inspiration to me

Preface

This book is written to provide a convenient and concise source of information on terminology for students in taxonomy, plant distribution, and speciation, especially in the higher plants (Pteridophyta and Spermatophyta). Its small size, compactness, and organization were designed to enable one easily to find the meaning of a term or to find a forgotten term with a minimum of effort.

About half of the book is given over to a glossary which is ample but intentionally not exhaustive. Obsolete and seldom used words have been omitted purposely.

The section on classified terms enables one quickly to review the terms relating to the subjects classified. This feature also makes it possible for one who is unable to recall a certain term to find it with little effort, and perhaps even more important, to find the appropriate word to use in his descriptions.

Many students of plant taxonomy become quite proficient in the rules and usages of the subject and readily recognize and call plants by their scientific names in the field, but the number who can tell the meaning of the scientific names is more limited. To aid in this respect, a generous section on specific epithets and their meanings has been added, and the author hopes that its use will broaden the users' understanding and add to their satisfaction.

Many graduate students of botany have studied neither Latin nor Greek. To encourage these students in an understanding of the word structure of botanical terms and names, more than six hundred word components derived from Greek and Latin have been included. The meaning of each is given with an example of a word in which it is used. Study of this section will provide an excellent foundation for the interpretation of terms in general and offer an aid in the selection and formation of new names and epithets when needed.

The laws, theories, and hypotheses in the Appendix all have a direct or indirect bearing on taxonomy, distribution, or speciation. They have been gleaned from a wide range of literature and brought together here with the hope that they may help the student broaden his concepts of natural phenomena in these fields of discipline.

I wish to acknowledge with grateful appreciation the valuable assistance of the following persons in particular: Dr. A. C. Andrews, University of Miami, Florida; Dr. H. M. Griffin, Oklahoma A. and M. College; and Prof. Robert Stratton, Oklahoma A. and M. College.

H. I. FEATHERLY

Stillwater, Oklahoma
August, 1953

Table of Contents

Glossary of Botanical Technical Terms

The following abbreviations will be used to indicate parts of speech: a. — adjective, n. — noun, v. — verb.

A

Abaxial – a. Said of an embryo which is out of the axis of the seed as the result of one-sided development of the albumen, or of the side of a lateral organ away from the axis.

Aberrant – a. Differing from the type of a species, genus, or higher taxon in one or more characters, but not readily assignable to another taxon.

Abortion – n. The arrested development of an organ.

Abortive – a. Imperfectly developed, not fully developed at maturity, as abortive stamens with filaments only.

Abrupt – a. Changing suddenly rather than gradually, as in a pinnately compound leaf without a terminal leaflet.

Absciss-layer – n. A layer of separation, especially with reference to the phenomena of defoliation.

Acantha – n. Thorn, spine, prickle.

Acarpic – a. Without fruit.

Acarpotropic – a. Not throwing off its fruits.

Acaulescent – a. Stemless or apparently stemless.

Acclimation – n. The process of becoming inured to a climate at first harmful.

Acclimatization – n. See Acclimation.

Accrescent – a. Enlarging with age, as the budscales of some hickories or sepals of some flowers.

Accumbent – a. Lying against another body.

Accumbent cotyledons – n. Cotyledons with edges lying against the radicle.

Acephalous – a. Headless.

Aceriform – a. Like a maple leaf.

Acerose – a. Needle-shaped; having a sharp rigid point, as the leaf of the pine.

Acetose – a. Acetic, sour.

Acicular – a. Slenderly needle-shaped.

Achene (akene) – n. A small, dry, indehiscent, one-seeded fruit in which the ovary wall is free from the seed.

Achenodium – n. A double achene, as the cremocarp of *Umbelliferae*.
Achilary – a. Without a lip, as in some orchids.
Achlamydeous – a. Without calyx or corolla, as in willows.
Acicula – n. The bristle continuation of the rachilla of a grass; a needle-like spine.
Acies – n. The edge or angle of certain stems.
Acquired character – n. A nonheritable environmental variation.
Acropetal – a. Produced in a succession toward the apex, as applied to development of organs (the antithesis of basipetal).
Acrophilous – a. Dwelling in the alpine region.
Actinomorphic – a. With radial symmetry.
Aculeate – a. Prickley; beset with prickles or sharp points.
Aculeolate – a. Beset with small prickles.
Acuminate – a. Tapering to a prolonged point; attenuate.
Acute – a. Distinctly and sharply pointed, but not drawn out.
Acyclic – a. Said of flowers whose parts are arranged spirally, not in whorls.
Adaxial – a. With side or face next to the axis; ventral.
Adenophorous – a. Gland-bearing.
Adherent – a. Attached or joined, though naturally or normally separate; adnate.
Adhesion – n. The union or fusion of unlike parts.
Adhesive disc – n. The disclike tip of some tendrils such as is found on Virginia Creeper.
Adnate – a. With unlike parts congenitally grown together.
Aduncate – a. Hooked.
Adventitious – a. Said of plants recently introduced, or of organs arising from abnormal positions, as buds from a root or roots from the stem or leaf.
Adventive – a. Imperfectly naturalized.
Adynamogyny – n. Loss of function in the female organ of a flower.
Aelophilous – a. Disseminated by wind.
Aestival – a. Belonging or peculiar to summer.
Aestivation – n. The arrangement of the parts of a flower in the bud.
Affinity – n. The closeness of relationship between plants as shown by similarity of important organs.
Afoliate – a. Aphyllous, leafless.
Agad – n. A beach plant.
Agamandroecism – n. In composites, the state of having male and neuter flowers in the same individual.
Agamogynaecism – n. In composites, the state of having female and neuter flowers in the same individual.
Agamohermaphroditism – n. A condition in which hermaphrodite and neuter flowers appear on the same plant.
Agamospermy – n. Seed production without fertilization.
Age and area hypothesis – n. The older the species the greater the area occupied.
Agglomerate – a. Crowded into a dense cluster, but not cohering.
Agglutinate – a. Stuck together, as the pollen-masses of asclepiads or orchids.
Aggregate – a. Assembled; collected together.
Aggregate fruit – n. A cluster of ripened ovaries traceable to separate pistils of the same flower and inserted on a common receptacle.
Aggregate species – n. A superspecies which may be compounded of more than one true species.
Agonisis – n. Certation; competition, as between pollen grains of different genotypes, in the rapid-

ity with which they can grow down the style.

Agrophilous – a. Growing in grain fields.

Agrostography – n. The description of grasses.

Agrostology – n. The study of grasses — their description, identification, classification, distribution, and habitat.

Agrotype – n. An agricultural race.

Agynic – a. Said of stamens which are free from the ovary; pistils wanting; destitute of pistils.

Aianthous – a. Constantly flowering; having everlasting flowers.

Aigialophilous – a. Beach-dwelling.

Aigicolous – a. Inhabiting a stony strand or beach.

Akene – n. See Achene.

Ala – n. A wing, a lateral petal of a papilionaceous flower; a membranous expansion of any kind, as in the seed of *Bignoniaceae*; the outer segment of the corolla lobes in some asclepiads.

Alate – a. Winged.

Albinism – n. The absence of pigmentation in organisms normally pigmented.

Albino – n. Any animal with congenital deficiency of pigment in skin, hair, eyes, etc; a plant with colorless chromatophores, due to the absence of chloroplasts or undeveloped chromoplasts.

Albinotic – a. Affected with albinism.

Albumen – n. Any deposit of nutritive material within the seed coats and not in the embryo.

Albuminous – a. Having albumen. See also Exalbuminous.

Aliferous – a. Having wings.

Allautogamia – n. The state of having two methods of pollination, one usual, and the other facultative.

Alliaceous – a. With the smell or taste of garlic; pertaining to the genus *Allium*.

Allochronic species – pl. n. Species which do not belong to the same time level, as opposed to contemporary, or synchronic.

Allogamous – a. Reproducing by cross-fertilization.

Allogamy – n. The pollination of a flower with pollen from another flower. See also Geitonogamy; Xenogamy.

Allogenous flora – n. Relic plants of an earlier prevailing flora and environment; epibiotic plants.

Allopatric – a. Inhabiting distinct separate areas.

Allotropous flower – n. A flower so shaped that its nectar is easily available to insects.

Alluring glands – pl. n. Glands in the pitchers of pitcher plants which tempt insects down the tube.

Alpestrine – a. Pertaining to the Alps or high mountains.

Alpine – a. Pertaining to the Alps or to the Arctic zone of a mountain; above timberline.

Alsad – n. A grove plant.

Alternate – a. Any arrangement of leaves or other parts not opposite or whorled; placed singly at different heights on the axis or stem.

Alveolate – a. With pits or depressions suggesting honey-comb.

Alveolation – n. A honey-combed condition.

Amathicolous – a. Growing on sandy plains.

Amathophilous – a. Growing in sandy plains or in sandy hills.

Ambiparous – a. Producing two kinds, as a bud which contains both flowers and leaves.

Ament – n. A catkin, a spike of flowers usually bracteate, pendulous, and deciduous.

Amentiferous – a. Bearing aments.

Amentum – n. Catkin.

Amethystine – a. Violet-colored.

Amixia – n. Cross-sterility.

Ammochthad – n. A sand-bank plant.

Ammophilous – a. Sand-loving.

Amphibious – a. Capable of living on land or in water.

Amphicarpous – a. Producing two kinds of fruit.

Amphichromy – n. The abnormal production of two different colors of flowers on the same stem.

Amphigean – a. Native around the world.

Amphimixis – n. Cross-fertilization. tion.

Amphitropous – a. Turned both ways; applied to an ovule with hilum intermediate between the micropyle and chalaza.

Amphora – n. The lower part of a pyxis, as in henbane.

Amplectant – a. Embracing, clasping by the base.

Amplexicaul – a. Clasping or embracing the stem, as a leaf.

Ampliate – a. Enlarged.

Ampulla – n. The flasks found on aquatics such as *Utricularia*.

Anadromous – a. Said of ferns in which the first set of nerves in each segment of the frond is given off on the upper side of the midrib toward the apex, as in *Aspidium* and *Asplenium*.

Anametadromous – a. Said of ferns in which the weaker pinnules are anadromous and the stronger are catadromous.

Anastomosing – a. Netted; interveined; said of leaves marked by cross-veins forming a network; interlacing.

Anastomosis – n. The union of one vein with another, the connection forming a reticulation.

Anatropous – a. The ovule reversed with micropyle close to the side of the hylum and the chalaza at the opposite end.

Ancipital – a. Two-edged.

Ancophilous – a. Loving mountain glens or valleys.

Androecium – n. The stamens of a flower (a collective term).

Androgynous – a. Hermaphoditic; having both male and female flowers in the same inflorescence. (Occasionally used with meaning of monoecious.)

Androphore – n. A support or column on which the stamens are raised.

Anemochore – n. An organism that is disseminated by the wind.

Anemochorous – a. Distributed by wind.

Anemogamous – a. Wind-pollinated.

Anemophilous – a. Said of flowers pollinated by wind.

Anemotropism – n. The tropic response of organisms to wind and air currents.

Angiosperm – n. A plant with seeds enclosed in an ovary or pericarp.

Angiospermous – a. Having the seeds borne within a pericarp.

Annotinous – a. A year old, or in yearly growths.

Annual – a. Of one year's duration; completing its life cycle in one year.

Annular – a. Said of any organs disposed in a circle.

Annulus – a. In ferns, the elastic organ which partially invests the theca, and at maturity bursts it; in *Equisetaceae*, the imperfectly developed foliar sheath below the fruit spike; the fleshy rim of the corolla in *Asclepiads*, as the genus *Stapelia*.

Anomalous – a. Not equal; unlike its allies in certain points; contrary to rule; unusual; out of the ordinary.

Anterior – a. Front; on the front side; away from the axis; toward the subtending bract.

Anthecology – n. The study of the flower and its environment.

Anthela - n. The panicle of *Juncus*, in which the lateral axes exceed the main axis.
Anthelate - a. With elongate flower-bearing branches, as in some junci.
Anthemia - n. See Anthemy.
Anthemy - n. A flower-cluster of any kind.
Anther - n. That portion of the stamen which bears the pollen.
Antheridium - n. In cryptogams, an organ or receptacle in which male sex cells are produced.
Antheriferous - a. Anther-bearing.
Antheroid - a. Anther-like.
Antherozoid - n. A male motile cell provided with cilia and produced in an antheridium, a sperm cell.
Anthesis - n. The act of flowering; strictly, the time of expansion of a flower, but often used to designate the flowering period.
Anthesmotaxis - n. The arrangement of the different parts of a flower.
Anthocarpous - a. Said of fruits with accessories, sometimes termed pseudocarps, as the strawberry and pineapple.
Anthracine - a. Coal-black.
Anthropochorous - a. Distributed by the action of man.
Anthropophilous - a. Said of plants which follow man, or cultivation.
Antrorse - a. Directed upward or forward; opposite of retrorse.
Apetalous - a. Without petals, or with a single perianth.
Aphaptotropism - n. The state of not being influenced by touching stems or other surfaces.
Apheliotropism - n. The act of turning away from the sun; negative phototropism.
Aphercotropism - n. The act of turning away from an obstruction.
Aphototropism - n. The act of turning away from light.
Aphyllous - a. Without leaves.
Apical - a. Pertaining to the apex or tip.
Apicula - n. A short, sharp, but not stiff point.
Apiculate - a. Having a minute pointed tip.
Apiculation - n. A short, sharp, but not stiff point, in which a leaf, petal, or other organ may end.
Apocarpous - a. With carpels separate, not united. See also Syncarpous.
Apocarpy - n. The condition of having the carpels separate.
Apogamous - a. Developed without fertilization, parthenogenetic.
Apomixy - n. The phenomenon of limited or no cross-fertilization. See also Panmixy.
Apophysis - n. An enlargement or swelling of the surface of an organ; the part of a cone scale that is exposed when the cone is closed.
Appendiculate - a. Furnished with an appendage.
Applanate - a. Flattened.
Appressed - a. Lying flat against an organ.
Approximate - a. Drawn close together, but not united.
Apterous - a. Wingless.
Apyrene - a. Said of fruit which is seedless.
Aquatic - a. Living in water.
Aquila - n. Eagle.
Arachnoid - a. Cobwebby; composed of soft, slender entangled hairs; spider-like.
Araneose - a. Like a spider-web.
Arboreous - a. Treelike or pertaining to trees.
Arborescent - a. Attaining the size or character of a tree; treelike.
Arbuscula - n. A small shrub with the aspect of a tree.
Archegonium - n. The organ or receptacle in which the female sex cells are produced in the higher cryptogams and some gymnosperms.

Arctic-alpine – a. Used for plants of arctic and alpine distribution but found only south of the Arctic zone.

Arcuate – a. Moderately curved; bent like a bow; descriptive of leaf venation of *Cornus, Ceanothus*, etc.

Arenaceous – a. Of or pertaining to sand; sandy; growing in sand.

Arenicolous – a. Growing in sand or sandy places.

Areolate – a. Marked with areoles, divided into distinct spaces; reticulate.

Areole – n. A space marked out on a surface.

Argenteoguttate – a. With silvery spots.

Argillaceous – a. Clayey, pertaining to clay, or clay-colored.

Argillicolous – a. Dwelling on clay.

Argute – a. Sharp.

Argyroneurous – a. With silver-colored nerves or veins.

Aril – n. An appendage or an outer covering of a seed growing out from the hilum or funiculus; sometimes it appears as pulpy covering.

Arillate – a. With an aril.

Aristate – a. Awned; provided with a bristle at the end, rarely on the back or edge.

Armed – a. Provided with any kind of strong and sharp defense, as of thorns, spines, prickles, barbs, etc.

Aromatic – a. Fragrant, spicy, pungent.

Arroyo – n. A water course, especially when dry (Southwestern U.S.).

Article – n. A segment of a constricted pod or fruit, as in *Desmodium*.

Articulate – a. Jointed; provided with nodes or joints, or places where separation may naturally take place.

Arundinaceous – a. Reedlike, having a culm like tall grasses.

Ascending – a. Rising up; produced somewhat obliquely or indirectly upward.

Asepalous – a. Without sepals.

Asexual – a. Sexless; without sex.

Asperous – a. Rough or harsh to the touch.

Assumentum (pl. **assumenta**) – n. The valve of a silique.

Assurgent – a. Ascending, rising.

Astigmatic – a. Wind-pollinated plants which do not possess stigmas, such as gymnosperms.

Asyngamic – a. Unable to cross by reason of differences in time of flowering.

Atavism – n. Ancestral resemblance, reversion to a more primitive type.

Atavistic form. – n. A reversion to the primitive form.

Atratous – a. Turning black; blackened, as in some species of *Carex*, the apex of the glumes being darkened.

Attenuate – a. Long tapering, acuminate.

Aurantiaceous – a. Orange-colored; like an orange.

Auricle – n. An ear; applied to earlike lobes at base of leaf blades and to small lobes at the summit of sheath in many species of *Gramineae*.

Auriculate – a. With earlike appendages.

Austral – a. Southern; occasionally applied to plants which are native to warmer countries, even if not from the Southern Hemisphere.

Autocarp – n. A fruit obtained as a result of self-fertilization.

Autogamous – a. Self-fertilized.

Autogamy – n. The fertilization of a flower by its own pollen, as in an autophilous flower.

Autoörthotropism – n. The tendency of an organ to grow in a straight line forward.

Autophilous – a. Self-pollinated.
Autophytic – a. Said of a plant able to produce its own food through the presence of chlorophyll.
Autumnal – a. Of or pertaining to autumn; flowering in autumn; serotinal.
Auxiliary – a. Helping.
Awl-shaped – a. Narrow and sharp-pointed; gradually tapering from base to a slender or stiff point.
Awn – n. A bristle-like appendage, especially on the glumes of grasses.
Axil – n. The upper angle formed between the axis and any organ that arises from it.
Axile – a. In the axis, said ordinarily of the placentae in the ovary.
Axillary – a. Situated in the axil.
Axis – n. The main or central line of development of any plant or organ; the main stem.
Azure – a. Sky blue.

B

Baccate – a. Berry-like; pulpy or fleshy.
Badious – a. Dark reddish-brown, chestnut-brown.
Balausta – n. The fruit of pomegranate with firm rind, berried within, crowned with the lobes of an adnate calyx.
Balsamiferous – a. Balsam-bearing.
Banner – n. The topmost petal in the corolla of a member of the pea family; standard; vexillum.
Barbed – a. With rigid points or short bristles, usually reflexed like the barb of a fishhook.
Barbellate – a. Finely barbed.
Barbulate – a. Finely bearded.
Barotropism – n. The response of an organism to changes in barometric pressure.
Barrier – n. Any obstacle that limits the distribution of a species; any condition that reduces or prevents crossbreeding.
Basifixed – a. Attached or fixed by the base, as an ovule that is affixed to its support by its bottom rather than by its side.
Basinerved – a. Veined from the base.
Basipetal – a. Growing in the direction of the base (the antithesis of acropetal).
Basonym – n. The specific or subspecific epithet which has priority and is retained when transferred to a new position.
Bast – n. Phloem; fibrous tissues serving for mechanical support.
Bay – a. Reddish-brown.
Beak – n. A long, prominent, and substantial projection; applied particularly to a prolongation of a fruit or carpel.
Beaked – a. Ending in a firm, prolonged, slender tip.
Beard – n. A long awn, or bristle-like hair.
Bearded – a. Bearing or furnished with long or stiff hairs.
Bellying – a. Swelling on one side, as in the corolla of many *Labiatae*.
Berry – n. Any simple fruit having a pulpy or fleshy pericarp, as the grape, gooseberry, tomato, or banana.
Betaceous – a. Of the beet; beet like.
Bicarpellary – a. Composed of two carpels.
Bicolored – a. Two-colored.
Bicruris – a. Two-legged, as the pollen masses of asclepiads.
Bicuspidate – a. Having two sharp points.
Bidentate. – a. Having two teeth.
Biennial – a. Of two seasons' duration from seed to maturity and death.
Biferous – a. Producing two crops of fruit in one season.

Bifid – a. Forked.
Bifurcate – a. Forked or two-pronged.
Bijugous – a. Yoked, two together.
Bilabiate – a. Two-lipped.
Bilateral – a. Arranged on opposite sides.
Bilobate – a. With two lobes.
Bilocular – a. Two-celled, with two compartments.
Binomial – a. The generic and specific name of an organism.
Biological races or species – pl. n. Races or species which differ only in their physiological behavior, being morphologically identical.
Biosystematy – n. Taxonomic studies involving cytology and genetics.
Biotype – n. A group of individuals all of one genotype.
Bipinnate – a. A condition in which both primary and secondary divisions of a leaf are pinnate.
Bisexual – a. Having both sexes on the same individual; a hermaphrodite.
Bivalvular – a. With two valves.
Bladdery – a. Inflated; empty, with thin walls like the bladder of an animal.
Blade – n. Lamina; the expanded portion of a leaf or petal.
Blastochore – n. A plant distributed by offshoots.
Bloom – n. The white, waxy, or pruinose covering on many fruits, leaves, and stems.
Blossom – n. A flower, especially of fruit trees.
Bole – n. The main trunk of a tree.
Bolochore – n. A plant distributed by propulsion.
Boreal – a. Northern.
Boss – n. A knoblike or rounded protuberance; umbo.
Bossed – a. With a rounded surface having a projection in its center.
Brachiate – a. Spreading with branches suggesting arms.
Bract – n. A modified leaf subtending a flower or belonging to an inflorescence.
Bracteate – a. With bracts.
Bracteody – n. The replacement of the floral whorls by bracts.
Bracteolate – a. With small bracts or bractlets.
Bracteole – n. A bractlet, or small bract.
Bracteose – a. Having conspicuous or numerous bracts.
Bractlet – n. Bract borne on a secondary axis, as on the peduncle or even on a petiole.
Bradycarpic – a. Fruiting after the winter, in the second season after flowering.
Bradyspore – n. A plant which disperses its seed slowly.
Branch – n. A lateral division of the stem, or axis of growth.
Branchlet – n. The ultimate divisions of a branch.
Bristle – n. A stiff hair.
Bristly – a. Bearing stiff, strong hairs.
Brotochore – n. A plant dispersed by man.
Brunescent – a. Brownish; becoming brown.
Brusque variation – n. A sudden, heritable deviation from type; mutation.
Bud – n. An embryonic axis with its appendages.
Bulb – n. A modified bud, usually underground; **imbricated** - with scaly modifications of the leaves, as in the lily; **tunicated** - with complete enveloping coats, as in the onion.
Bulbiferous – a. Bulb-bearing.
Bulbil – n. A bulb arising from the mother bulb.
Bulblet – n. A little bulb produced in the leaf axils, inflorescence, or other unusual places.
Bulbose – a. Having bulbs or the structure of a bulb.

Bulbous – a. Having the character of a bulb.

Bullate – a. Blistered or puckered on the surface, as the leaf of a Savoy cabbage.

Bulliform – a. Applied to large thin-walled epidermal cells of most *Gramineae* and *Cyperaceae*.

Bumble-bee flowers – pl. n. See Humble-bee flowers.

Bur, burr – n. Any rough or prickly envelope, as of a pericarp, a persistent calyx, or an involucre; any plant which bears burs.

Bursicle – n. A pouchlike receptacle.

Bursicule, bursicula – n. The pouchlike expansion of the stigma into which the caudicle of some orchids is inserted.

Bush – n. A low shrub, branching from the ground.

C

Caducous – a. Falling off early, or prematurely, as the sepals in some plants.

Caerulescent – a. Bluish; becoming blue.

Caesp'tose – a. Growing in tufts.

Calathiform – a. Cup-shaped.

Calcarate – a. Spurred.

Calcareous – a. Of or pertaining to calcium carbonate (limestone), as a calcareous soil.

Calceiform – a. Shoe-shaped.

Calcicolous – a. Growing best in a soil with a high lime content.

Callosity – n. A hardened thickening.

Callous – a. Having the texture of a callus.

Callus – n. A hard prominence or protuberance; in a cutting or on a severed or injured part, the roll of new covering tissue; an extension of the flowering-glume below its point of insertion and grown to the axis or rachilla of the spikelet.

Calycanthemy – n. Petalody of the calyx; the formation of colored petal-like structures in place of a normal calyx.

Calyciflorous – a. Having calyx, corolla, and stamens adnate.

Calyculate – a. Calyx-like; bearing a part resembling a calyx; particularly, furnished with bracts against or underneath the calyx resembling a supplementary or outer calyx.

Calyptra – n. A hood or lid; particularly, the hood or cap of the capsule of a moss or lid in the fruit of *Eucalyptus*.

Calyx – n. The outermost circle of the floral envelopes.

Cambium – n. A layer, usually regarded as one cell thick, of persistent meristematic tissue (referring to vascular and cork cambia); or a persistent meristematic layer which gives rise to secondary wood and secondary phloem (Vascular cambium).

Campanula – n. Small bell.

Campanulate – a. Bell-shaped.

Campestrian – a. Of plains or open country.

Camptodromous – a. Said of venation in which the secondary veins curve towards the margins, but do not form loops.

Campylodromous – a. Said of venation with its primary veins curved in a more or less bowed form towards the leaf apex.

Campylotropous – a. Said of an ovule or seed which is curved in its formation so as to bring the micropyle or true apex down near the hilum.

Canaliculate – a. Longitudinally channeled.

Cancellate – a. Latticed; resembling lattice-work.

Candelabra hairs – n. Stellate hairs in two or more tiers.

Canescence – n. Hoariness, usually with gray pubescence.

Canescent – a. Becoming hoary, usually with a gray pubescence.
Cantharophilous – a. Said of plants that are pollinated by beetles.
Capillary – a. Hairlike; very slender.
Capitate – a. Headed; in heads; formed like a head; aggregated into a very dense or compact cluster.
Capoe – n. A palm thicket (Brazil).
Capreolate – a. Having tendrils.
Capsella – n. A small seed vessel.
Capsular – a. Pertaining to a capsule; formed like a capsule.
Capsule – n. A simple dry fruit, the product of a compound pistil splitting along two or more lines of suture.
Cardinal – a. Of cardinal-red color.
Carina – n. A keel; used either for the two combined lower petals of a papilionaceous flower or for a salient longitudinal projection on the center of the lower surface of an organ, as on the lemmas of many grasses.
Carneous – a. Flesh-colored.
Carpel – n. A simple pistil; one unit of a compound pistil; in conifers, the cone scale of the female cone.
Carpellate – a. Possessing carpels.
Carphospore – n. A plant whose seeds are disseminated by means of a scaly or chaffy pappus.
Carpography – n. Description of fruits.
Carpophore – n. A portion of receptacle prolonged between the carpels, as in *Umbelliferae*.
Caruncle – n. An excrescence or appendage at or about the hilum of the seed.
Carunculate – a. With a caruncle.
Caryopsis – n. The grain or fruit of most grasses, with the seed coat grown fast to the pericarp.
Castaneous – a. Chestnut-colored; dark brown.
Catadromous – a. Said of ferns in which the first set of nerves in each segment of the frond is given off on the basal side of the midrib, as in *Osmunda*.
Catkin – n. A flexible, usually pendulous scaly spike bearing apetalous, unisexual flowers; ament.
Caudate – a. With a tail or tail-like appendage.
Caudex. – n. The woody base of a perennial plant.
Caudicle – n. A cartilaginous strap which connects certain pollen-masses to the stigma, as in orchids.
Caulescent – a. More or less stemmed or stem-bearing; having an evident stem above ground.
Cauline – a. Pertaining or belonging to the stem.
Caulis (pl. **Caules**) – n. The stalk or stem of a plant.
Cecidium – n. A gall produced by fungi or insects, in consequence of infection; an abnormal growth.
Cell – n. Any structure containing a cavity, as the cell of an anther or ovary; locule; a unit of plant structure.
Cellular – a. Pertaining to cells.
Cement-disk – n. The retinaculum in orchids.
Cenanthy – n. Suppression of the stamens and pistil, leaving the perianth empty.
Censer-action – n. The action of capsules that, like censers (incense-burners), partially open by valves, the seeds being gradually shaken out by the wind, as in *Papaver* and *Cerastium*.
Centrifugal – a. In inflorescences, blooming from the inside outward, or from top to base.
Centripetal – a. In inflorescences, blooming from the outside inward, or from the base upward.
Centrospore – n. A plant with spurred fruits.
Centrum – n. The central portion, as the large central air space in hollow stems, as in *Equisetum*.

Cerasiferous – a. Cherry-bearing.

Cereal – n. Any grass whose seeds serve as food (from Ceres, the goddess of agriculture).

Ceriferous – a. Wax-bearing; waxy.

Cernuous – a. Drooping; inclining somewhat from the perpendicular; nodding.

Certation – n. Competition, as between pollen grains of different genotypes, in the rapidity with which they can grow down the style; agonisis.

Cespitose, caespitose – a. Matted; growing in tufts; in little dense clumps; said of low plants that make tufts or turf of their basal growths.

Chaff – n. Small membranous scales, degenerate bracts in many *Compositae*; the outer envelopes of cereal grains.

Chalaza – n. That part of the ovule or seed in which the nucellus joins the integuments; the base of the nucellus, always opposite the upper end of the cotyledons.

Chalicad – n. A gravel slide plant.

Chalicophilous – a. Dwelling in gravel slides.

Channeled – a. Grooved longitudinally.

Chartaceous – a. Having the texture of writing paper.

Chasmogamous – a. With pollination taking place while the flower is open (the opposite of cleistogamous).

Chasmogamy – n. The opening of the perianth at flowering time (the opposite of cleistogamy).

Chasmophilous – a. Having a fondness for crannies.

Chemotropism – n. Curvature in response to a chemical stimulus.

Cheradad – n. A wet sandbar plant.

Cheradophilous – a. Loving dry habitats; dwelling in dry places.

Chersad – n. A plant of a dry waste.

Chersophilous – a. Dwelling in dry places.

Chionad – n. A snow-plant.

Chionic – a. Of snow fields.

Chiropterophilous – a. Said of plants which are pollinated by bats.

Chledocolous – a. Dwelling in waste places.

Chledophilous – a. Preferring waste places.

Chloranthous – a. Having green, usually inconspicuous flowers.

Chloranthy – n. The reversion of petals to green leaves.

Chlorophyll – n. The green coloring matter in the cells of autophytic plants.

Chlorophyllous – a. Containing chlorophyll.

Chlorosis – n. A yellowing of the plant due to chlorophyll deficiency.

Chlorotic – a. Lacking chlorophyll.

Choripetalous – a. Polypetalous, with petals separate.

Chorisis – n. Separation of an organ (leaf, petal, stamen, etc.) into more than one.

Chorology – n. The geographic study of the distribution of organisms.

Chromosome – n. One of the small bodies, ordinarily definite in number in the cells of a given species and often more or less definite in shape, into which the chromatin of the cell nucleus resolves itself previous to the mitotic division of the cell.

Chrysanthine – a. Yellow-flowered.

Chrysophyllous – a. Golden-leaved.

Cicatrice – n. Scar, the mark left by the separation of one from another, as by the leaf from the stem.

Cicatrix – n. See Cicatrice.

Ciliate – a. Said of a margin fringed with hairs.

Ciliolate – a. Said of a margin fringed with small hairs.

Cilium (pl. **cilia**) – n. Used generally in the plural to designate marginal hairs.

Cincinnus – n. A one-branched scorpoid cyme.
Cineraceous – a. Somewhat ashy in tint.
Cinereous – a. Ash-colored; light gray.
Circinate – a. Coiled from the top downward; coiled into a ring, or partially so.
Circumscissile – a. Opening or dehiscing along a horizontal line around the fruit or anther, the valve usually coming off like a lid.
Cirriferous – a. Curl-bearing, tendril-bearing.
Cirrus – n. A curl, a tendril.
Citreous – a. Lemon yellow.
Cladode – n. A branch of a single internode simulating a leaf; a cladophyll.
Cladophyll – n. A branch assuming the form and function of a leaf; a cladode.
Class – n. The name of the taxon which is next higher than order.
Clastotype – n. A fragment from the original type.
Clathrate – a. Latticed.
Clavate – a. Club-shaped; said of a long body thickened toward one end.
Clavellate – a. Dimminutive of clavate.
Clavicle – n. A tendril, cirrus.
Claviculate – a. Furnished with tendrils or hooks.
Claviform – a. Club-shaped.
Claw – n. The long narrow petiole-like base of the petals or sepals in some flowers; the modified auricle of some grass leaves, such as wheat and barley.
Cleft – a. Divided into lobes separated by narrow or acute sinuses which extend more than halfway to the midrib.
Cleistogamous – a. Having fertilization occur within the unopened flower.
Cleistogamy – n. The state of being cleistogamous.
Cleistogene – n. A plant which bears cleistogamous flowers.
Cleistogenous – a. Cleistogamous.
Cleistogeny – n. The state of bearing cleistogamous flowers.
Climbing – a. Ascending by using other objects as supports.
Clinandrium – n. The anther bed in orchids, that part of the column in which the anther is concealed.
Clinanthium – n. The receptacle in *Compositae*.
Cline – n. A series of form changes; a gradient of biotypes along an environmental transition.
Clinium – n. The receptacle of a composite flower.
Clip – n. The seizing mechanism in the flowers of asclepiads.
Clitochore – n. A plant that is distributed by falling or sliding.
Clockwise – a. In the same direction as the hands of a clock, dextrorse.
Clon – n. See Clone.
Clone – n. The vegetatively produced progeny of a single individual.
Close fertilization – n. Fertilization by its own pollen.
Coalescence – n. The union of like parts or organs.
Coarctate – a. Crowded together.
Cob – n. Rachis of the pistillate corn (maize) spike.
Coccus – n. A berry; in particular, one of the parts of a lobed fruit with one-seeded cells.
Cochlea – n. A closely coiled legume.
Cochlear – a. Spoon-shaped; said of a form of imbricate aestivation with one piece exterior.
Cochleate – a. Spiral, like a snail shell.
Coelospermous – a. Hollow-seeded; said of the seedlike carpels of *Umbelliferae*, with ventral face incurved at the top and bottom as in *Coriander*.

Coenocarpium - n. The collective fruit of an entire inflorescence, as a fig or pineapple.

Coenospecies - n. The total sum of possible combinations of a genotype compound; a variable hybrid of two Linneons or ecospecies.

Coerulescent - a. See Caerulescent.

Coherent - a. Two or more similar parts or organs joined.

Cohesion - n. Union of like parts.

Collar - n. The transition zone between primary stem and root; the back side of the union of the blade and sheath in grasses.

Collateral - a. Descriptive of accessory buds arranged on either side of a lateral bud.

Colliculose - a. Covered with little round elevations or hillocks.

Colonial - a. Forming colonies; used chiefly for plants with asexual reproduction.

Column - n. A combination of stamens and styles into a solid central body, as in orchids; the lower, twisted portion of an awn of grasses, not always present.

Coma - n. The hairs at the end of some seeds; the tuft at the summit of the inflorescence, as in the pineapple; the entire head of a tree.

Comal tuft - n. A tuft of leaves at tip of a branch.

Combinatio nova (comb. nov.) - n. New combination, i.e., a hitherto unpublished scientific plant name based on a rearrangement of name already published.

Comb-shaped - a. Pectinate.

Commissure - n. The place of joining or meeting, as the face by which one carpel joins another.

Comose - a. Bearing a tuft or tufts of hair.

Complanate - a. Flattened, compressed.

Complicate - a. Folded upon itself.

Compound - a. Similar parts aggregated into a common whole.

Compound inflorescence - n. An inflorescence composed of secondary ones.

Compound leaf - n. One leaf consisting of two or more blades (leaflets).

Compound pistil - n. Two or more carpels coalesced into one body.

Compressed - a. Flattened; especially, flattened laterally.

Concave - a. Hollow, as the inside of a saucer.

Concolor - a. Of the same color.

Conduplicate - a. Folded together lengthwise with the upper surface within, as in the blades of many grasses.

Cone - n. The fruit of a pine, cycad, or fir-tree with scales forming a strobile; an inflorescence or fruit with overlapping scales.

Conelet - n. The diminutive of cone, applied to a cone of the first year in hard pines.

Conferted - a. Closely packed, or crowded.

Confluent - a. Blended into one, passing by degrees one into another.

Congested - a. Crowded.

Conglomerate - a. Clustered.

Conical - a. Having the form of a cone, as the carrot.

Conifer - n. A cone-bearer.

Coniferous - a. Producing or bearing cones.

Conjugate - a. Coupled, or in pairs.

Connate - a. United congenitally or subsequently.

Connate-perfoliate - a. United at the base in pairs around the supporting axis.

Connivent - a. Coming together or converging but not organically connected.

Conocarpium - n. An aggregate fruit consisting of many fruits on a conical receptacle, as the strawberry.

Conoidal - a. Cone-shaped.

Conopodium – n. A conical floral receptacle.

Constipate – a. Crowded, or massed together.

Contorted – a. Twisted or bent; in aestivation, the same as convolute.

Contortuplicate – a. Twisted and plaited or folded; twisted back upon itself.

Contracted – a. Said of the inflorescences that are narrow and dense, the branches short or appressed.

Convergent – a. Applied to veins which run from the base to the apex of a leaf in a curved manner.

Convergent evolution – n. The evolution of similar structures produced by different means in different lines of descent.

Convex – a. Having a more or less rounded surface.

Convolute – a. Said of floral envelopes in the bud in which one edge overlaps the next part, as sepal or petal or lobe, while the other edge or margin is overlapped by a preceding part; rolled up from the sides longitudinally.

Copious – a. Abundant.

Coppice – n. A small wood which is regularly cut at stated intervals, the new growth arising from the stools.

Coracoid – a. Shaped like a crow's beak.

Cordate – a. Heart-shaped; said of leaves having the petiole at the broader and notched end.

Cordiform – a. Shaped like a heart.

Coriaceous – a. Like leather.

Cork – n. Protective tissue replacing the epidermis in older superficial parts of plants; the outer cells contain air, and are elastic and spongy in texture, but impervious to liquids.

Corm – n. A solid bulblike stem, usually subterranean, as the "bulb" of *Crocus*, or *Gladiolus*.

Cormatose – a. Producing corms.

Cormel – n. A corm arising from a mother corm.

Corneous – a. Horny, with a horny texture.

Cornet – n. A hollow hornlike growth.

Corniculate – a. Bearing or terminating in a small hornlike protuberance or process.

Corolla – n. The inner floral envelope, composed of separate or connate petals.

Corolline – a. Seated on a corolla; corolla-like; petaloid, or belonging to a corolla.

Corolloid – a. Corolline; corolla-like; petaloid.

Corona – n. Crown, coronet; any appendage or intrusion that stands between the corolla and stamens, or on the corolla, as the cup of a daffodil, or that is the outgrowth of the staminal part or circle, as in the milkweed.

Coronate – a. Crowned; with a corona.

Coroniform – a. Shaped like a crown.

Corrugate – a. Wrinkled.

Cortex – n. Rind or bark.

Cortical – a. Relating to bark.

Corymb – n. Short and broad, more or less flat-topped indeterminate flower cluster, the outer flowers opening first.

Corymbiform – a. Shaped like a corymb.

Corymbose – a. Arranged in corymbs.

Coryphad – n. An alpine meadow plant.

Costa – n. A rib, as a midrib.

Costate – a. Ribbed; with one or more longitudinal ribs or nerves.

Cotyledon – n. Seed leaf; the primary leaf or leaves in the embryo.

Cotype – n. An additional or associate type specimen from which a species is described.

Counterclockwise – a. Sinistrorse, turning the reverse way of clock-hands.

Crampon – n. A hook or adventitious root which acts as a support, as in ivy.

Craspedodromous – a. A condition in which the lateral veins of a leaf run from midrib to margin without dividing.

Crateriform – a. Saucer- or cup-shaped; shallow.

Creatospore – n. A plant with nut fruits.

Creeper – n. A trailing shoot that roots throughout most of its length; sometimes said of a tightly clinging vine.

Creeping – a. Running along on the ground and rooting.

Cremocarp – n. A dry, seedlike fruit composed of two one-seeded carpels invested by an epigynous calyx, separating when ripe into mericarps.

Crena – n. A rounded tooth or notch.

Crenad – n. A plant growing near a spring.

Crenate – a. Said of a margin with rounded or blunt teeth.

Crenicolous – a. Dwelling in brooks fed by springs.

Crenophilous – a. Dwelling near a spring.

Crenulate – a. Finely crenate.

Creophagous – a. Carnivorous, as applied to plants.

Crested – a. With elevated and irregular toothed ridge.

Crisp – a. Curled.

Cristate – a. Crested.

Cristulate – a. With small crests.

Cross-pollination – n. The pollination of the stigma by pollen derived from another plant not in the same clone.

Crown – n. Corona; the base of a tufted, herbaceous, perennial grass; the hard ring or zone at the summit of the lemma of some species of *Stipa*; the part of a stem at the surface of the ground; a part of a rhizome with a large bud, used in propagation.

Crosier – n. Any plant structure with a curled end, as the young leaves of most ferns.

Cruciate – a. Cross-shaped, said especially of the flowers of *Cruciferae*.

Crucifer – n. A plant with four petals and tetradynamous stamens; a member of the family *Cruciferae*.

Cruciform – a. Cross-shaped.

Crustaceous – a. Of hard and brittle texture.

Crymophilous – a. Dwelling in polar regions.

Cryotropism – n. Movement induced by cold or frost.

Cryptanthous – a. With hidden flowers; cleistogamous; the stamens remaining enclosed in the flower.

Ctenoid – a. Comblike, pectinate.

Cucullate – a. Hooded or hood-shaped.

Culm – n. The jointed stem of grasses and sedges.

Cultigen – n. Plant or group known only in cultivation; presumably originating under domestication; contrast with indigen.

Cultivor – n. A variety or race that has originated and persisted under cultivation, but not necessarily referable to a botanical species.

Cultrate – a. Having the shape of a knife blade.

Cuneate – a. Wedge-shaped; triangular, with the narrow end at the point of attachment, as of leaves or petals.
Cuneifoliate – a. With wedge-shaped leaves.
Cuneiform – a. Wedge-shaped.
Cup – n. An involucre, as of an acorn.
Cupule – n. The cup of such fruits as the acorn; an involucre composed of bracts adherent by their base at least.
Cupuliform – a. Cup- or cupule-shaped.
Cuspidate – a. Tipped with a sharp, rigid point.
Cutin – n. A substance present as a thin continuous external layer on the outer wall of the epidermis of a leaf or stem.
Cyamium – n. A kind of follicle resembling a legume.
Cyanochrous – a. Having a blue skin.
Cyanthiform – a. Cup-shaped.
Cyanthum – n. The ultimate inflorescence of *Euphorbia*, consisting of a cuplike involucre bearing the flowers from its base.
Cycle – n. A term used for one turn of a helix or spire, in leaf arrangement; for a whorl of floral envelopes.
Cyclic – a. Said of foliar structures arranged in whorls; coiled into a cycle or relating to a cycle.
Cylindrical – a. Elongated with a circular cross section.
Cyme – n. A broad, more or less flat-topped determinate flower-cluster, with central flowers blooming first.
Cymose – a. Cyme-like.
Cymule – n. A small cyme.
Cynarrhodium – n. A fruit like that of the rose, fleshy, hollow, and enclosing achenes, as a rose hip.
Cypsela – n. An achene invested by an adnate calyx, as the fruit of *Compositae*.

D

Dactyliferous – a. Finger-bearing.
Dasycarpous – a. Thick-fruited.
Dasyphyllous – a. Thick-leaved.
D. B. H. – n. Diameter breast-high.
Decamerous – a. In tens.
Decandrous – a. Having ten stamens.
Decapetalous – a. Having 10 petals.
Deciduous – a. Not persistent; said of leaves falling in autumn or of floral parts falling after anthesis.
Decompound – a. More than once compound.
Decumbent – a. Reclining or lying on the ground, but with the ends ascending.
Decurrent – a. Said of a leaf or leaf scar, part of which extends in a ridge down the twig below the point of insertion.
Decussate – a. In pairs alternately crossing at right angles.
Definite – a. Precise; of a certain number, as of stamens not exceeding twenty; applied to inflorescence, it means cymose.
Definite inflorescence – n. An inflorescence in which the axis terminates in a flower, cymose, determinate.
Deflexed – a. Bent or turned abruptly downward.
Deflorate – a. Past the flowering state.
Defoliation – n. The act of shedding leaves.
Dehiscence – n. The method or process of opening of a seed-pod or an anther.
Dehiscent – a. That which dehisces, as the opening of an anther or fruit along regular lines of suture.
Dehisce – v. To open spontaneously when ripe, as seed capsules.
Deliquescent – a. Dissolving or melting away; said of a stem which loses itself by repeated branching; opposed to excurrent.

Deltoid - a. Triangular, delta-like.

Deme - n. Any specified assemblage of taxonomically closely related individuals.

Dendrocolous - a. Dwelling on trees.

Dendroid - a. Treelike; shaped like a tree.

Dendrology - n. The study of trees — their description, classification, identification, and distribution.

Dendrophilous - a. Dwelling on or among trees; tree-loving.

Dentate - a. Said of a margin with sharp teeth pointing outward.

Denticulate - a. Minutely or finely dentate.

Dentoid - a. Tooth-shaped.

Denudate - a. Stripped, made bare, or naked.

Depauperate - a. Reduced or undeveloped, impoverished, dwarfed.

Depurlation - n. The act of throwing off bud-scales in leafing.

Deplanate - a. Flattened or expanded.

Depressed - a. More or less flattened endwise or from above; pressed down.

Derma (pl. **dermata**) - n. Surface of an organ, bark, rind, or skin.

Descending - a. Tending gradually downward; as the branches of some trees or as the roots.

Desmobrya - n. A group of ferns in which the fronds are adherent to the caudex.

Determinate - a. Said of an inflorescence in which the terminal flower blooms slightly in advance of its nearest associates; limited in number or extent.

Dextrorse - a. Turning to the right, clockwise.

Diadelphous - a. Said of stamens formed in two groups through the union of their filaments.

Diadromous - a. Said of a venation shaped like a fan, as in *Ginkgo biloba*.

Dialycarpic - a. Having a fruit composed of distinct carpels.

Dialypetalous - a. Polypetalous.

Diandrous - a. Possessing two stamens.

Dianthic - a. Fertilized by the pollen from the same plant.

Diaphototropism - n. The act of placing itself at right angles to incident light.

Diaphragm - n. A dividing membrane, or partition, as in the pith of *Juglans*.

Diaspore - n. A disseminule; any spore, seed, fruit, or other portion of a plant capable of producing a new plant.

Diatropism - n. The act of organs placing themselves crosswise to an operating stimulus.

Dicarpellary - a. Composed of two carpels.

Dichasium - n. A cyme with two lateral axes.

Dichlamydeous - a. Having double perianth, calyx and corolla.

Dichogamous - a. Hermaphrodite with one sex maturing earlier than the other, stamens and pistil not synchronizing.

Dichogamy - n. A condition in perfect flowers in which the sexes do not mature simultaneously.

Dichotomous - a. Branching by constantly forking in pairs.

Diclinism - n. The separation of the anther and stigma in space, as dichogamy is in time.

Diclinous - a. Having staminate and pistillate flowers either on the same plant or on different plants.

Dicotyledones - pl. n. A class of angiosperms differentiated by possession of two cotyledons.

Dicotyledonous - a. Having two cotyledons.

Dictyodromous - a. With reticulate venation.

Dicymose - a. Doubly cymose.

Dicyclic – a. Having a series of organs arranged in two whorls, as a perianth; biennial.

Didymous – a. Found in pairs, as the fruits of *Umbelliferae*; divided into two lobes.

Didynamous – a. Said of four-stamened flowers with stamens in pairs, two long, two short, as in some *Labiatae*.

Diffuse – a. Loosely branching or spreading; of open growth.

Digitate – a. Finger-like; compound with the members arising from one point, as the leaflets of horse chestnut.

Digonous – a. Two-angled, as the stems of some cacti.

Dimerous – a. Flowers with the parts in twos.

Dimidiate – a. Halved, as a condition in which half an organ is so much smaller than the other as to seem wanting.

Dimorphic –a. Occurring in two forms.

Dimorphous –a. Occurring in two forms.

Dioecious – a. Unisexual, the male and female elements in different plants.

Diphotic –a. With two surfaces equally lighted.

Diplobiont – n. A plant flowering or fruiting twice each season.

Dipterid – n. Fly flowers, visited chiefly by dipterous flies.

Dipterous – a. Two-winged.

Disarticulate – v. To separate at a joint, as the leaves in autumn.

Disc, disk – n. Development of the torus within the calyx or within the corolla and stamens; the central part of a capitulum in *Compositae* as opposed to the ray; the base of a pollinium; the expanded base of the style in *Umbelliferae*; in a bulb, the solid base of the stem around which the scales are arranged.

Disc flowers – n. The tubular flowers in the center of the heads of *Compositae*, as distinguished from the ray flowers.

Dischisma (pl. **dischismata**) – n. The fruit of *Platystemon*, which divides into longitudinal carpels, each of which again divides transversely.

Disciform – a. Depressed and circular like a disk.

Discoid – a. With a round thickened lamina and rounded margins.

Disepalous – a. With two sepals.

Disk – n. See Disc.

Disk flowers – n. See Disc flowers.

Dispermous – a. Two-seeded.

Dispersal – n. The act of dispersing or scattering.

Dissected – a. Deeply divided, or cut into many segments.

Dissemination – n. The act of dispersing or scattering such objects as seed, fruit, pollen, etc.

Disseminule –n. See Diaspore.

Dissepiment – n. A partition in an ovary or pericarp caused by the adhesion of the sides of the carpellary leaves.

Dissilient – a. Bursting asunder.

Distant – a. Said of similar parts not closely aggregated; opposed to approximate; remote.

Distichous – a. Conspicuously two-ranked, in two rows.

Distinct – a. Separate; not united with parts in the same series.

Diurnal – a. Occurring in the daytime; sometimes used meaning ephemeral.

Divaricate – a. Widely divergent.

Divergent – a. Inclining away from each other.

Divided – a. Characterized by a lobing or segmentation which extends to the base.

Dodecagynous – a. Possessing twelve pistils or distinct carpels.

Dodecamerous - a. In twelve parts, as in a cycle.

Dodecandrous - a. Normally possessing twelve stamens, occasionally extended to more than twelve.

Dolabriform - a. Axe-shaped or hatchet-shaped.

Doliform - a. Barrel-shaped.

Domesticated - a. Thriving under cultivation.

Dormant - a. Said of parts which are not in active life.

Dorsal - a. Relating to the back, or attached thereto; the surface turned away from the axis, which in a leaf is the lower surface; opposed to ventral.

Dorsifixed - a. Attached by the back.

Dorsiventral - a. With a distinct upper and lower surface.

Down - n. Soft pubescence; the pappus of such plants as thistles.

Drepaniform - a. Sickle-shaped.

Drepanium - n. A sickle-shaped cyme.

Drimyphilous - a. Salt-loving.

Driodad - n. A plant of a dry thicket.

Dromotropism - n. The irritability of climbing plants which results in their spiral growth.

Drupaceous - a. Resembling a drupe, possessing its character, or producing similar fruit.

Drupe - n. A fleshy one-seeded indehiscent fruit, with seed enclosed in a stony endocarp called a pit.

Drupelet -n. One drupe in a fruit made up of aggregate drupes, as in a raspberry.

Duct - n. A tube or canal which carries resin, latex, or oil.

Dubious - a. Doubtful, said of plants whose structure or affinities are doubtful.

Dumetose - a. Bushy; relating to bushes.

Dumose - a. Full of bushes, or of shrubby aspect.

Dysteleology - n. The supposition that nature (and especially organic evolution) lacks any foreordained direction or purpose.

Dystropous - a. Said of an insect whose visit is injurious to the flower.

E

Ebeneous - a. Black as ebony.

Ebracteate - a. Without bracts.

Eburneous - a. Ivory-white, white more or less tinged with yellow.

Ecad - n. A form arising by adaptation to environment.

Ecblastesis - n. The appearance of buds within a flower; proliferation of an inflorescence.

Echinate - a. Armed with prickles.

Echma (pl. **echmata**)-n. The hardened hook-shaped funiculus which supports the seed in most *Acanthaceae.*

Ecological - a. Pertaining to the relation of organisms to their environment.

Ecology - n. The study of organisms in relation to their environment.

Ecospecies - n. A habitat form of a species.

Ecotype - n. A habitat type of plant.

Ectopy - n. The abnormal position of an organ.

Edaphotropism - n. Tropic responses to the soil.

Edoble - n. A plant whose seeds are scattered by propulsion through turgescence.

Eeltrap hairs - pl. n. Hairs which detain insect visitors, as in *Sarracenia* and *Aristolochia.*

Efflorescence - n. The season of flowering, anthesis.

Effuse - a. Patulous, expanded, loosly spreading.

Eglandular - a. Without glands.

Elater – n. In *Equisetum*, four club-shaped hygroscopic bands attached to the spores, which serve for dispersal.

Ellipsoid – n. An elliptic solid.

Elliptic – n. A flat part or body that is oval and narrowed to rounded ends.

Elliptical – a. Shaped like an ellipse, oblong with rounded ends.

Elongate – a. Stretched; lengthened.

Emasculation – n. The removal of the anthers from a bud or flower.

Emarginate – a. With shallow notch at the apex.

Embracing – a. Clasping by the base, amplectant.

Embryo – n. The rudimentary plant formed in the seed.

Emersed – a. Raised above and out of the water.

Enaulophilous – a. Dwelling in sand draws.

Endemic – a. Indigenous or native to.

Endocarp – n. The inner layer of a pericarp.

Endogenous – a. Growing by internal accessions, as monocotyledonous stems.

Endosperm – n. The albumen of a seed in angiosperms; in gymnosperms the prothallium within the embryo sac.

Enneanderous – a. With nine stamens.

Ensiform – a. Sword-shaped, as in the leaf of the *Iris*.

Entire – a. Without toothing or division, with even margin.

Entomogamous – a. Insect-pollinated.

Entomogamy – n. The pollination of flowers by insects.

Entomophilous – a. Said of a plant whose flowers are pollinated by insects.

Envelope – n. The surrounding part.

Environment – n. The aggregate of surrounding conditions.

Epharmonic convergence – n. Morphological and anatomical similarity between taxonomically unrelated, or distantly related plants.

Epharmosis – n. Organic adaptation to a changed environment.

Ephemer – n. A flower that closes after a short term of expansion.

Ephemeral – a. Persisting for one day only, as flowers of spiderwort.

Epibiotic species – n. Endemic species that are relics of a past flora and compose a minor portion of the biota of most regions.

Epicalyx – n. A series of bracts close to and resembling the calyx.

Epicarp – n. The external layer of a pericarp, exocarp.

Epicormic – a. Said of adventitious buds which develop on the trunks of trees, and of branches which develop on the body of a forest tree from which surrounding trees have been removed.

Epiderm – n. The true cellular skin or covering of a plant below the cuticle.

Epigeotropism – n. Tropism resulting in growth on the surface of the soil.

Epigeous – a. Growing upon or above the ground; on land as opposed to water; said of seedlings which bring their cotyledons above ground.

Epigynous – a. Borne on the ovary; said of floral parts in which the ovary is inferior and not perigynous.

Epipetalous – a. Borne upon the petals; placed before the petals.

Epiphyte – n. A plant which grows on other plants, but not parasitically.

Epithet - n. A single descriptive word or single descriptive phrase; in taxonomy, it is applied to the subdivisions of genera, to the second component of the name of species, and to the subdivisions of species.

Equal - a. Alike as to length, size, or number.

Equitant - a. Folded over as if astride; used for conduplicate in which the leaves are folded together lengthwise in two ranks, as in *Iris*.

Eremophilous - a. Desert-loving.

Ericoid - a. Erica-like, like the heath.

Eriophorous - a. Wool-bearing, densely cottony.

Erose - a. Irregularly toothed or eroded as though bitten or gnawed.

Erubescent - a. Blushing red.

Erythrosorus - a. With red sori.

Esculent - a. Suitable for human food, edible.

Essential - a. The necessary constituent of an existing object.

Essential organs - pl. n. Organs which are absolutely necessary, as stamens and pistil.

Estipitate - a. Without a stipe.

Estival - a. Of or pertaining to summer. See also Aestival.

Ethnobotany - n. The study of the relations between man (especially primitive man) and his surrounding vegetation.

Etiolated - a. Blanched.

Eutropic - a. Twining in the direction of the sun, clockwise, dextrorse; said of flowers to which only a restricted class of specialized insects can gain access.

Evergreen - a. Remaining green during its dormant season; said of plants that are green throughout the year.

Exalbuminous - a. Destitute of albumin, used only of seed in which the embryo occupies the whole cavity within the testa.

Exannulate - a. Said of ferns which do not possess an elastic ring around their sporangia.

Exasperate - a. Rough, with hard projecting points.

Excentric - a. One-sided, off-center, abaxial.

Excurrent - a. Running through to the apex and beyond, as a mucro; a stem that remains central, the other parts being regularly disposed around it.

Exfoliate - v. To come off in scales or flakes, as the bark of the sycamore.

Exfoliating - a. Coming off in thin layers.

Exine - n. The outer coat of a pollen grain.

Exocarp - n. The outer layer of a pericarp.

Exogenous - a. Said of growth by the addition of layers on the outside, as with dicotyledons.

Exotic - a. Foreign, not native, from another region.

Explanate - a. Spread out flat.

Explosive speciation - n. The rapid production, within one locality, of a number of new species from a single species.

Exserted - a. Sticking out; projecting beyond, as stamens from a perianth; not included.

Exsiccata (sing. **exsiccatum**) - pl. n. Dried plants, usually in sets for sale or for subscribers.

Exsiccated - a. Dried.

Exstipulate - a. Without stipules.

Extravaginal - a. Beyond or outside the sheath; said of branches springing from buds, which break through the sheath of the subtending leaf, chiefly in grasses.

Extrorse – a. Facing or opening outward.

Exuviae – pl. n. Cast-off parts, as shed scales.

F

Facultative – a. Having the power to live under different conditions, as a facultative parasite, a plant which is normally saprophytic, but which may exist wholly or in part as a parasite; opposed to obligate.

Falcate – a. Sickle- or scythe-shaped.

Family – n. The taxon next higher than the genus.

Farinaceous – a. Mealy, like meal or pretaining to meal.

Farinose – a. Covered with a meal.

Fasciated – a. Much flattened; an abnormal widening and flattening of the stem as though several stems had coalesced in one plane.

Fascicle – n. A close cluster or bundle of flowers, leaves, stems, or roots.

Fascicled – a. In clusters or bundles.

Fasciculate – a. In close bundles or clusters.

Fastigiate – a. Parallel, clustered, and erect, as the branches of *Populus fastigiata.*

Faucal – a. Pertaining to the throat of a gamopetalous corolla.

Fauces – n. The throat of a gamopetalous corolla.

Faveolate – a. Honey-combed, alveolate.

Favose – a. Honey-combed, as the receptacles of many *Compositae.*

Feather-veined – a. With secondary veins proceeding from the midrib; penninerved.

Fecundation – n. Fertilization.

Felted – a. Matted with intertwined hairs.

Female – n. The fruiting element in plants, the pistil and its analogues.

Fenestra – n. Lattice, window, an opening through a membrane.

Fenestrate – a. Pierced with holes, as the septum in some *Cruciferae.*

Feral – a. Wild; not cultivated.

Fertile – a. Said of pollen-bearing stamens and seed-bearing fruits; capable of producing fruit.

Ferruginous – a. Rust-colored.

Fertile flowers – pl. n. Female flowers, those which possess pistils.

Fetid – a. Having a disagreeable odor.

Fiber, fibre – n. A thread, or threadlike structure; a long, slender, thick-walled cell, as in sclerenchyma; the fusiform cells of the inner bark; the ultimate rootlets.

Fibriform – a. Fiber-shaped.

Fibril, fibrilla – n. A small fiber.

Fibrillose – a. Furnished or abounding with fine fibers.

Fibrous – a. Having numerous woody fibers, as the rind of a coconut.

Fibro-vascular – a. Composed of woody fibers, xylem vessels, and sieve tubes.

Ficoid – a. Figlike.

Fiddle-shaped – a. Panduriform.

Filament – n. The part of a stamen that supports the anther; threadlike structures.

Filamentous – a. Formed of filaments or fibers.

Filical – a. Fernlike, relating to the *Filicineae.*

Filicoid – a. Fernlike.

Filiferous – a. With filaments or threads.

Filiform – a. Threadlike, long and very slender.

Filipendulous – a. Hanging from a thread.

Fimbria – n. A fringe.

Fimbriate – a. Fringed, the hairs longer or coarser as compared with ciliate.

Fimbrillate – a. With a minute fringe.
Fimbriolate – a. Very finely fimbriate.
Fimetarious – a. Growing on or among dung.
Fimicolous – a. Growing on manure.
Fissile – a. Tending to split, or easily split.
Fistula – n. A pipe, or hollow cylinder.
Fistular – a. Hollow-cylindrical.
Fistulous – a. Hollow-cylindrical, as the leaf and stem of an onion.
Flabellate – a. Fan-shaped, dilated in a wedge-shape, sometimes plaited.
Flabelliform – a. Fan-shaped.
Flaccid – a. Withered and limp, flabby.
Flagelliform – a. Whip-shaped.
Flask – n. The utricle of *Carex*.
Flattening – n. The fasciation of a stem; the production of a cladodium.
Flavescent – a. Yellowish, becoming yellow.
Fleshy – a. Succulent.
Flexible – a. Capable of being bent, but elastic enough to be able to resume its original shape.
Flexuous – a. Bent alternately in different directions.
Floccose – a. Bearing tufts of woolly hairs.
Flocculent – a. The diminutive of floccose.
Flocculose – a. Bearing small tufts of woolly hairs.
Flora – n. The aggregate of plants of a country or district, or a work which contains the enumeration of them.
Floral – a. Of or pertaining to flowers.
Floral diagram – n. A drawing to show the relative position and number of the constituent parts.
Floral envelope – n. The perianth leaves, calyx, and corolla.
Floral formula – n. A formula composed of letters, figures, and symbols arranged to show number, union, sex, shape, elevation, etc. of the floral parts.
Floral glume – n. The lower glume of the floret in grasses; the lemma.
Florepleno – a. With full or double flowers.
Florescence – n. Anthesis, the peroid of flowering.
Floret – n. The lemma and palea with included flower (stamens and pistil); a small flower; one of a cluster, as in *Compositae*.
Floricane – n. The flowering cane, usually the second year's development of the primocane, in *Rubus* and other such genera.
Floriferous – a. Flower-bearing.
Flos (pl. **flores**) – n. a perfect flower with some protecting envelope.
Floscule – n. A little flower, a floret.
Floss – n. The down of certain *Compositae*, as thistle-down.
Flower – n. (See Flos.) A modified plant structure concerned with the production of seeds in the angiosperms.
Flower bud – n. An unexpanded flower, as distinct from a leaf-bud or mixed bud.
Flower-glume – n. The lower of the two scales which subtend the flower of grasses, the lemma.
Fluminal – a. Said of a plant which grows in running water.
Fluted – a. Regularly marked by alternating ridges and groovelike depressions.
Fluvial – a. Applied to plants growing in streams.
Foliaceous – a. Leaflike; said particularly of sepals, calyx-lobes, and of bracts that in texture, size, or color look like leaves.

Foliage – n. The leafy covering, especially of trees.

Foliar – a. Pertaining to a leaf.

Foliose – a. Closely clothed with leaves, leafy.

Follicle – n. A single carpellate dry fruit dehiscing along one line of suture.

Follicular – a. Of or pertaining to a follicle.

Foramen (pl. **foramina**) – n. An aperture, especially that in the outer integuments of the ovule; micropyle.

Foraminose – a. Perforated by holes.

Forcipate – a. Forked liked pincers.

Forest – n. Land covered with trees exclusively, or with undergrowth of shrubs or herbs.

Fornicate – a. Provided with arched scalelike appendages in the corolla tube, as in *Myosotis*.

Fornix (pl. **fornices**) – n. A small arched scale.

Fovea – n. A depression or pit, as in the upper surface of the leaf-base in *Isoëtes*, which contains the sporangium; the seat of the pollinium in orchids.

Foveolate – a. Marked with small pittings.

Free – a. Not joined to another organ.

Frond – n. The foliage of ferns and some other cryptogams; the leaves of palms.

Frondose – a. Leafy; frondlike or bearing fronds.

Fructescence – n. The time of maturity of fruit.

Fructiferous – a. Producing or bearing fruit.

Fructification – n. The act of fruiting.

Fruit – n. A mature ovary or ovaries with or without closely related parts.

Fruit dots – n. The sori of ferns.

Frustraneous – a. Said of *Compositae* with disk flowers hermaphrodite, and those of the ray neuter or imperfect.

Frutescent – a. Becoming shrubby.

Frutex – n. A woody plant destitute of a trunk.

Frutical – a. Shrubby with a soft, woody stem, such as shrubby species of *Geranium*.

Fruticose – a. Shrubby or shrublike in the sense of being woody.

Fugacious – a. Soon falling or fading; not permanent.

Fulcrum (pl. **fulcra**) – n. An appendage of leaves, as a prickle, tendril, stipule, etc.

Fuliginous – a. Sooty, or sootcolored.

Fulvous – a. Yellow, tawny.

Funicle – n. A stem or thread which connects the ovule or seed to the placenta; funiculus.

Funiculus – n. A stem or thread which connects the ovule or seed to the placenta; funicle.

Funnel – n. A space below the thick outer coats of the macrospore in *Marsiliaceae*, into which the apical papilla projects.

Funnelform – a. With tube gradually widening upward and passing insensibly into the limb, as in many flowers of *Convolvulus*; infundibuliform.

Furcate – a. Forked with terminal lobes which are like prongs.

Furrowed – a. With longitudinal channels or grooves; sulcate; striate on a large scale.

Fuscous – a. Grayish-brown.

Fusiform – a. Spindle-shaped; narrowed both ways from a swollen middle, as Dahlia roots.

G

Galbulus – n. The fruit of the *Cypress*, a modified spherical cone, the apex of each carpellary scale being enlarged and somewhat fleshy.

Galea – n. A petal shaped like a helmet placed next to the axis, as in *Aconitum*.

Galeate – a. Hollow and vaulted, as in many labiate corollas.

Galericulate – a. Covered, as with a hat.

Galochrous –a. Milk-white.

Gamopetalous – a. With corolla of one piece; petals united.

Gamophyllous – a. With leaves united by their edges.

Gamosepalous – a. With sepals united into one piece.

Geitonogamy – n. Pollination by pollen from another flower on the same plant.

Gelatinous – a. Jelly-like.

Geminate – a. In pairs, binate, twin.

Gemma (pl. **gemmae**) – n. A bud or a body analogous to a bud capable of producing a new plant.

Gemmiferous – a. Bearing buds.

Gemmiparous – a. Bearing gemmae.

Gene – n. A unit of inheritance, which occupies a fixed place on a chromosome.

Gene-flow – n. The spread of genes which takes place within a group (variety, subspecies, or species) as a result of outcrossing followed by natural crossing within the group.

Generitype – n. The type species of a genus.

Geniculate – a. Abruptly bent so as to resemble the knee-joint, as of awns and the lower nodes of some culms.

Genitalia – pl. n. The stamens and pistils; reproductive organs.

Genospecies – n. A group, all the members of which are genotypically identical.

Genotype – n. The type of a genus, the species upon which the genus was established.

Gens – n. A tribe in botany.

Genus – n. The smallest natural group containing distinct species; large genera are frequently divided for the sake of convenience into subgenera, but the generic name is applied to all species.

Geocarpy – n. The subterraneous ripening of fruits which have developed from a flower above ground.

Geodiatropism – n. The tendency to place an organ at right angles to the force of gravity.

Geographic speciation – n. The gradual formation of new species by reason of spatial isolation of different stocks of the original species.

Geonasty – n. The act of curving toward the earth.

Geophilous – a. Earth-loving; said of plants which fruit underground.

Geotaxis – n. Orientation of organisms with reference to gravity.

Gibbous – a. Swollen on one side as the glume in *Sacciolepis*; a pouchlike enlargement of the base of an organ, as of a calyx.

Glabrate – a. Nearly glabrous, or becoming glabrous with maturity or age.

Glabrous – a. Smooth, devoid of pubescence or hair of whatsoever form.

Gladiate – a. Flat, straight, or slightly curved, with acute apex and approximately parallel edges, ensiform, swordlike.

Gland – n. An acorn, or acorn-like fruit; a definite secreting structure on the surface, embedded, or ending a hair; any protuberance of like nature which may not secrete, as the warty swellings at the base of the leaf in the cherry and peach.

Glandular – a. Furnished with glands, or of the nature of glands.

Glandule – n. A viscid gland in orchids and asclepiads, which holds the pollen-masses in their place; the retinaculum.

Glanduliferous – a. Bearing glands.

Glandulose – a. Glandular.

Glans (pl. **glandes**) – n. A fruit one-seeded by abortion, or a few-seeded, dry, inferior, indehiscent pericarp seated within a cupular involucre, as the nut of an acorn as distinguished from the cup.

Glareose – a. Frequenting gravel.

Glaucescent – a. Becoming sea-green; somewhat glaucous.

Glaucophyllous – a. Glaucous-leaved.

Glaucous – a. Covered with a "bloom" or a whitish substance that rubs off, as of a plum or cabbage leaf.

Glittering – a. With luster from a polished surface which is not uniform.

Globose – a. Spherical, globular.

Globular – a. Spherical.

Glochid – n. A barbed hair or bristle.

Glochideous – a. Pubescent with barbed bristles.

Glochidiate – a. Pubescent with barbed bristles.

Glome – n. A rounded head of flowers, as *Echinops*.

Glomerate – a. In a dense, compact cluster or clusters.

Glomerule – n. A cluster of capitula in a common involucre.

Glumaceous – a. With glumes; resembling the glumes of grasses.

Glume – n. The chaffy two-ranked members of the inflorescence of grasses and similar plants; one of the two empty bracts at the base of a grass spikelet.

Glume, empty – n. Glume which subtends a spikelet, and does not include a flower.

Glume, flowering – n. The glume in grasses which includes a flower; the lemma.

Glutinous – a. Covered with a sticky exudation.

Gnaurs – pl. n. Burls or knotty excrescences on tree trunks or roots, probably from clusters of adventitious buds; burls.

Gnesiogamy – n. Fertilization between different individuals of the same species.

Gorge – n. The throat of a flower.

Gossypine – a. Cottony, flocculent, like the hairs on the seeds of *Gossypium*.

Gourd – n. A fleshy, one-celled, many-seeded fruit, with parietal placentation, as a melon; a pepo.

Gramineous – a. Relating to grass, or grain-bearing plants.

Graminicolous – a. Grass-inhabiting.

Granular – a. Covered with very small grains; minutely or finely mealy.

Granulose – a. Granular.

Gray – a. A cold, neutral tint.

Gregarious – a. Growing in groups or masses, but not matted.

Grumose – a. Crumby.

Guttation – n. The act of forming drops.

Gymnospermous – a. Bearing naked seeds.

Gynandrous – a. With the stamens adnate to the pistil, as in orchids.

Gynecandrous – a. With staminate and pistillate flowers in the same spike, the pistillate at the apex; used chiefly in reference to the *Cyperaceae*.

Gynobase – n. An enlargement or prolongation of the receptacle bearing the ovary.

Gynodioecious – a. Dioecious, with some flowers hermaphrodite, others pistillate only on separate plants.

Gynoecium – n. The pistil or pistils of a flower; the female part of a flower.

Gynomonoecious – a. With pistillate and perfect flowers on one plant.

Gynophore – n. Stipe of an ovary prolonged within the calyx.

Gynospore – n. One of the larger reproductive bodies (female) in the *Isoëtaceae.*

Gynostegium – n. A sheath or covering of the gynoecium of whatever nature.

Gynostemium – n. The compound structure resulting from the union of the stamens and pistil in *Orchidaceae.*

Gypsophilous – a. Growing on gypsum soils.

H

Habit – n. The general appearance of a plant, whether erect, prostrate, climbing, etc.

Habitat – n. The kind of locality in which a plant grows.

Hair – n. An outgrowth of the epidermus consisting of one to several cells.

Halberd-shaped – a. Hastate; sagittate (arrow-shaped), with the lobes turned out.

Halophilous – a. Salt-loving.

Halophyte – n. A plant which grows in saline soil.

Halophytic – a. Growing in saline soil.

Hamate – a. Hooked at the tip.

Hamous – a. Hooked.

Haplocaulous – a. Having a simple unbranched stem.

Haplochlamydeous – a. Monochlamydeous, having a single perianth.

Haplostemonous – a. Possessing only one whorl of stamens.

Hastate – a. Halberd-shaped, sagittate, with the basal lobes turned outward.

Haustorium – n. A sucker of ectoparasitic plants.

Head – n. A dense spherical or flat-topped inflorescence of sessile flowers clustered on a common receptacle.

Heart-shaped – a. Cordate; broadly ovate with two rounded lobes at the base.

Hebecarpous – a. Having a fruit covered with downy pubescence.

Helad – n. A marsh plant.

Heliad – n. A heliophyte, or sun-loving plant.

Helicoid – a. Curved or spiraled like a snail-shell.

Heliotropic – a. Turning in response to sunlight.

Heliotropism – n. The act of turning in response to the sun.

Helohylophilous – a. Dwelling in wet forests.

Hemeranthous – a. Day-flowering.

Hemicarp – n. A half-carpel, a mericarp.

Hepaticous – a. Liver-colored, dark purplish-red.

Herb – n. A plant naturally dying to the ground at the end of the growing season, without persistent stem above ground, and lacking definite woody, firm structure.

Herbaceous – a. Not woody; dying to the ground each year; said also of soft branches before they become woody.

Hercogamy – n. The condition of a hermaphrodite flower in which some structural peculiarity prevents self-fertilization, requiring some other form of pollination.

Hermaphrodite – a. With stamens and pistil in the same flower.

Hesperidium – n. A superior, polycarpellary, syncarpous berry, pulpy within, and externally covered with a tough rind.

Heterandrous – a. With two sets of stamens; said of flowers whose stamens vary in size or length.

Heterocarpous – a. Producing more than one kind of fruit.

Heterocephalous – a. With staminate and pistillate flowers on separate heads on the same plant.

Heterochlamydeous – a. With the perianth clearly divided into a calyx and a corolla.

Heterodistyly – n. Dimorphism, the presence of two forms of flowers, one with long and the other with short styles.

Heterodromous – a. Having spirals of changing directions, as in some tendrils, or phyllotaxis.

Heterodromy – n. A condition in which two spirals take different or opposite courses.

Heteroecious – a. Existing on different hosts.

Heterogameon – n. A species made up of races which, if selfed, produce morphologically stable populations, but when crossed may produce several types of viable and fertile offspring.

Heterogamous – a. Bearing two kinds of flowers, as in *Compositae,* the floret of the ray may be neuter or unisexual, and those of the disk hermaphrodite; an abnormal arrangement of the sexual organs.

Heterogamy – n. The state of being heterogamous.

Heterogeneous – a. Not uniform in kind.

Heterophyllous – a. With leaves of more than one shape.

Heterostyly – n. The state of having unlike or different length styles.

Hexagynia – n. A Linnean order of plants possessing six pistils.

Hexagynous – a. With six pistils.

Hexamerous – a. With parts in sixes.

Hexandria – n. A Linnean class characterized by the possession of six stamens.

Hexandrous – a. With six stamens.

Hexapetalous – a. With six petals.

Hexaphyllous – a. With six leaves.

Hibernaculum (pl. **hibernacula**) – n. The winter resting part of a plant, as a bud or underground stem.

Hibernation – n. Passing the winter in a dormant state.

Hibernal – a. Relating to winter.

Hiemal – a. Relating to winter.

Hilum – n. The scar or mark on a seed indicating the point of attachment.

Hinoideous – a. With the veins proceeding from the midrib parallel and undivided.

Hip – n. The fruit of the rose; technically, a cynarrhodion.

Hippocrepiform – a. Horseshoe-shaped.

Hirsute – a. With stiff or bristly hairs.

Hirsutulous – a. Slightly hirsute.

Hirtellous – a. Softly or minutely hirsute or hairy.

Hispid – a. Beset with rough hairs or bristles.

Hispidulous – a. Somewhat or minutely hispid.

Hoary – a. Covered with a close white or whitish pubescence.

Holosericeus – a. Covered with a fine and silky pubescence.

Holotype – n. The single specimen chosen as the basis for the original diagnosis.

Homalotropous – a. Said of organs which grow in a horizontal direction.

Homocephalic – a. Term used by Delpino for homogamy in which the pollen of one flower fertilizes the stigma of another flower in the same inflorescence.

Homoclinous – a. Term used by Delpino for homogamy in which the pollen of one flower fertilizes the stigma of the same complete flower.

Homodromous – a. Having the spirals all of the same direction.

Homoeandrous – a. Having uniform stamens.

Homogamous – a. Characterized by homogamy.

Homogamy–n. Simultaneous ripening of pollen and stigmas in a perfect flower.

Homogeneous – a. All of one kind, uniform.

Homogony – n. With the pistils and stamens of all flowers of uniform length.

Homolepidous – a. With one kind of scales.

Homologous – a. Of one type; of similar origin though varying in form and function.

Homonym – n. A name having the same spelling as another name in the same language but different in meaning; in taxonomy, homonyms are two or more names having the same spelling but applied to two or more taxa of the same rank based upon different types. A later homonym is illegitimate.

Homostyly – n. The state of having similar or equally long styles.

Homotropic – a. Fertilized by pollen from the same flower.

Homotropous – a. Curved or turned in one direction; said of the embryo of an anatropous seed, with the radical next to the hilum.

Hospitator – n. A plant which shelters ants.

Host – n. A plant which nourishes a parasite.

Hover-fly flowers – pl. n. Those adapted for pollination by *Syrphidae*.

Humble-bee (bumble-bee) flowers – pl. n. Flowers especially adapted for the visits of the species of *Bombus*.

Humifuse – a. Spreading over the ground.

Husk – n. The outer covering of certain fruits or seeds.

Hyacinthine – a. Dark purplish-blue; hyacinth-like in habit; a scape bearing spicate flowers.

Hyalescent – a. Somewhat hyaline or becoming so.

Hyaline – a. Thin and translucent or transparent.

Hybrid – n. A cross usually between two species of the same genus.

Hybrid swarm – n. A hybrid population typically occurring in the overlapping area between interfertile species or subspecies.

Hydrocarpic – a. Said of aquatic plants whose flowers are pollinated above the water but withdrawn below the surface for development, as in *Vallisneria*.

Hydrochore – n. A plant distributed by water.

Hydrophilous – a. Dwelling in wet places or water; pollinated by water.

Hydrophyte – n. A water plant, partially or wholly immersed.

Hydrotropism – n. The act of turning in response to the influence of water.

Hyemal – a. Incorrect variation of hiemal.

Hygrochastic – a. Said of plants in which the bursting of the fruit and the dispersion of seeds are caused by absorption of water.

Hygrophilous – a. Said of moisture-loving plants.

Hygroscopic – a. Altering form or position through changes in humidity.

Hylacolous – a. Tree-dwelling.

Hylodophilous – a. Dwelling in dry woods.

Hylophilous – a. Dwelling in forests.

Hymenosepalous – a. With membranous sepals.

Hypanthium – n. An enlargement or development of the torus under the calyx.

Hyperboreal – a. Of the far North.

Hyperdromy – n. A condition in which anadromous and catadromous venation occurs on one side of a fern-frond.

Hyphodromous – a. Said of a condition in which the veins are sunk in the substance of the leaf, and thus are not readily visible.

Hypocarpogenous – a. Having flowers and fruit placed underground.

Hypochil – n. The (often fleshy or otherwise modified) basal portion of the labellum or lip in *Orchidaceae.*

Hypocotyl – n. The axis of an embryo below the cotyledons, but not passing beyond them.

Hypocrateriform–a. Salver-shaped, with a salver-shaped corolla.

Hypogeous – a. Under the earth or soil.

Hypogynium – n. The perianth-like structure subtending the ovary in *Seleria* and some other *Cyperaceae.*

Hypogynous – a. Free from, but inserted beneath, the pistil or gynoecium.

Hypogyny – n. The condition of possessing hypogynous flowers.

Hyponastic – a. Said of a dorsiventral organ in which the ventral surface grows more actively than the dorsal, as shown in flower expansion; employed by Van Tieghem for anatropous or campylotropous ovules when the curvature is in an upward direction.

Hyponym – n. A name to be rejected for want of an identified type.

Hysteranthous – a. Said of leaves which are produced after the flowers, as in the almond.

I

Icon (pl. **icones**) – n. A plate, engraving, picture, or other form of image; usually used in the plural in botany as in *icones plantarum.*

Icosahedral – a. Having twenty sides, as the pollen grains of *Tragopogon.*

Icosandrous – a. With twenty or more stamens.

Imbricated – a. Overlapping as shingles on a roof; in aestivation, said of a calyx or corolla in which one piece must be wholly internal and one wholly external, or overlapping at the edge only.

Immersed – a. Entirely under water; embedded in the substance of a leaf.

Immobile – a. Immovable, as many anthers (opposed to versatile).

Imparipinnate – a. Unequally pinnate, odd pinnate, with a single terminal leaflet.

Imperfect – a. Said of a flower with one of the sexes wanting.

Implexed – a. Entangled, interlaced.

Implicated – a. Entangled, interwoven.

Impressed – a. Bent inward, hollowed, or furrowed as if by pressure.

Incanescent – a. Becoming gray, canescent.

Incarnate – a. Flesh-colored.

Incised – a. Cut sharply and irregularly, more or less deeply.

Included – a. Not protruding beyond the surrounding organ; not exserted.

Incompatibility – n. The inability of pollen to effect fertilization.

Incompletae – pl. n. Usually synonymous with *Monochlamydeae,* but variously circumscribed by different authors.

Incomplete – a. Said of flowers with one or more of the four floral organs wanting.

Incrassate - a. Made thick or stout, as the leaves of houseleek.

Incubous - a. With the leaves inserted obliquely so that the base of each is covered by the upper portion of the next lower, as in *Bazzania.*

Incumbent - a. Resting or leaning upon.

Incumbent anther - n. An anther attached to the inner face of its filament.

Incumbent cotyledon - n. A cotyledon with its back lying against the radical.

Indefinite - a. Uncertain or not positive in character; too many for easy enumeration, as abundance of stamens; in an inflorescence, indeterminate.

Indefinite growth - n. Continuous growth until cold weather kills back the immature terminal bud as well as the outer end of the twig as in certain species of *Rubus.*

Indefinite inflorescence - n. An inflorescence that is indeterminate or centrifugal, acropetal according to some authors; one that blooms progressively from outside inward or from the bottom upward.

Indehiscent - a. Not opening by valves or along regular lines.

Indeterminate - a. Said of an inflorescence in which the flowers open progressively from the base upward or from the outside inward.

Indigen - n. A native, not introduced.

Indigenous - a. Native to the country, not introduced.

Indigo - n. A deep blackish-blue obtained from various species of *Indigofera.*

Indument - n. Any hairy covering or pubescence.

Induplicate - a. With the margins bent inwards, and the external face of these edges applied to each other without twisting.

Indurated - a. Hardened.

Indusial - a. Pertaining to indusia.

Indusial flaps - n. A false indusium as in *Woodwardia.*

Indusiate - a. Possessing an indusium.

Indusium (pl. **indusia**) - n. The epidermal outgrowth covering the sori or "fruit-dots" on ferns; a ring of collecting hairs below the stigma.

Induviae - pl. n. Persistent portions of the perianth or leaves which wither but do not fall off; scale-leaves.

Induviate - a. Clothed with withered remnants.

Inequilateral-a. Asymmetrical, unsymmetrical.

Inermous - a. Without spines or prickles, unarmed.

Inferals - n. A division of gamopetalous dicotyledons proposed for *Rubiaceae, Compositae, Campanulaceae,* etc.

Inferior - a. Said of one organ when below another, as an inferior ovary with an adnate or superior calyx.

Inferior ovary - n. Ovary with the perianth located on top.

Inflated - a. Bladder-like, swollen, puffed up.

Inflexed - a. Turned in at the margins.

Inflorescence - n. Mode of flower-bearing; technically less correct but much more common in the sense of a flower-cluster.

Inflorescence, definite - n. A condition in which each axis in turn is terminated with a flower, as in a cyme.

Inflorescence, indefinite - n. A condition in which the floral axis is capable of continuous extension, as in a raceme.

Infra-axillary – a. Below the axil, sub-axillary.
Infundibuliform–a. Funnel-shaped.
Infundibular – a. Funnel-shaped.
Innate – a. Borne at the apex of the supporting part, as some anthers.
Innovation – n. The basal shoot of a perennial grass; a newly formed shoot which becomes independent from the parent stem by dying off behind.
Inrolled – a. Rolled inwards.
Insectivorous – a. Said of those plants which capture insects and presumably absorb nutriment from them.
Insect pollination – n. The transfer of pollen from the anther to the stigma by insects; entomophily.
Inserted – a. Attached to, or growing out of.
Insular – a. Pertaining to an island.
Integument – n. The covering of an organ or body; the envelope of an ovule.
Intercalary inflorescence – n. A condition in which the main axis continues to grow vegetatively after giving rise to the flowers.
Intercostal – a. Between the ribs or nerves of a leaf.
Interfoliaceous – a. Between the leaves of a pair, as the stipules of many *Rubiaceae*.
Internerves – pl. n. The space between the nerves.
Internode – n. The part of the stem between two successive nodes.
Interrupted – a. Having any symmetrical arrangement destroyed; not continuous.
Interruptedly pinnate – n. Pinnate with small leaflets interposed with those of larger size.
Intramarginal – a. Within and near the margin.
Introduced – a. Brought from another region.
Introrse – a. Turned or faced inward or toward the axis, as an anther facing toward the center of the flower.
Intumescent – a. Tumid, swollen, enlarged, distended.
Invaginated – a. Enclosed in a sheath.
Inverted – a. Turned over; end-for-end; top side down.
Involucel – n. A secondary involucre; a small involucre about the parts of a cluster.
Involucellate – a. With a secondary involucre.
Involucral – a. Pertaining to an involucre.
Involucrate – a. With an involucre.
Involucre – n. A cluster of bracts subtending a flower or inflorescence.
Involute – a. Rolled in from the edges, the upper surface within.
Irregular – a. Wanting in regularity of form; asymmetric, as a flower which cannot be halved in any plane, or one that is capable of bisection in one plane only, zygomorphic.
Irregular flower – n. A flower with some parts different from other parts in the same series.
Isadelphous – a. Characterized by equal brotherhood, the number of stamens in each group being equal.
Isogenous – a. Having the same origin.
Isolation – n. The separation of one group from another within a species so that crossing between the groups is prevented.
Isomerous – a. With the members of successive circles of equal numbers.
Isotype – n. A specimen believed to be a duplicate of the holotype.
Iteology – n. The study of the genus *Salix*, willows.

J

Joint – n. An articulation, as a node in grasses or other plants.

Jointed – a. With nodes or points of real or apparent articulation.

Jordanon – n. A microspecies; a small species of slight variability.

Jugum – n. A pair of leaflets; the ridges on the fruits of *Umbelliferae.*

Julaceous – a. Bearing catkins, amentaceous.

Juvenile – a. Young, early forms.

Juxtaposition–n. The relative position in which organs are placed; a placing or being placed side by side.

K

Keel – n. A central dorsal ridge; the united petals of a papilionaceous flower.

Kernel – n. The nucellus of an ovule, or of a seed, that is, the whole body within the coats.

Kettle-trap – n. A flower such as that of *Aristolochia,* which imprisons insects until fertilization is effected.

Key – n. A short statement of the contrasted characters of a genus or other group.

Key fruit – n. The samara of ash.

Kingdom – n. One of the highest groups of organic nature; the *Vegetable Kingdom* includes all plants.

Kleistogamous – a. See Cleistogamous.

Klone – n. See Clone.

Knee – n. An abrupt bend in a stem or tree-trunk; an outgrowth of some tree-roots.

Knight-Darwin Law – n. "That no organic being fertilizes itself for an eternity of generations"; "Nature abhors perpetual self-fertilization."

L

Label – n. The pinnule or ultimate segment of a fern-frond; labellum.

Labellum – n. The third petal of orchids, usually enlarged, and by torsion of the ovary having become anterior from its normal posterior position; a similar petal in other flowers; lip.

Labiate – a. Lipped; a member of the *Labiatae.*

Labium – n. The lower lip of a labiate flower; the lip subtending the ligule in *Isoëtes.*

Labyrinthiform – a. With complicated sinuous lines or winding passages.

Lacerate – a. Torn at the edge or irregularly cleft, as in some ligules.

Laciniate – a. Cut into lobes separated by deep, narrow, irregular incisions.

Lactiferous – a. Latex-bearing.

Lacuna – n. An air space in the midst of tissue; said of the vallecular canals of *Equisetum*; a hole or cavity.

Lacuno-rugose – a. Marked with irregular wrinkles, as the stone of a peach.

Lacustrine – a. Belonging to, or inhabiting lakes or ponds.

Ladaniferous – a. Ladanum-bearing.

Laevigate – a. Smooth, as if polished.

Lagenarious – a. Of a bottle or flask.

Lamella – n. A thin plate.

Lamellate – a. See Lamellose.

Lamellose – a. Having thin plates.

Lamina – n. The limb, blade, or expanded part of a leaf or petal.

Lanate – a. Clothed with woolly and interwoven hairs.

Lanceolate – a. Lance-shaped, rather narrow, tapering to both ends with the broadest part below the middle.

Lanose – a. Woolly.

Lanuginous – a. Cottony or woolly.

Lapidose – a. Growing amongst stones.

Lasiocarpous – a. With pubescent fruit.

Lateral – a. On or at the side.

Laterally compressed – a. Flattened from the sides as certain spikelets, glumes, and lemmas.

Latericious – a. Brick-red.

Latex – n. The milky juice of such plants as spurge or milkweed; the moisture of the stigma.

Latifoliate – a. Broad-leaved.

Latisquamate – a. Broad-scaled.

Laurine – a. Laurel-like.

Lax – a. Loose, distant.

Layering – a. Rooting, said of side branches.

Leader – n. The primary or terminal shoot of a tree.

Leaf – n. A lateral projection on a stem at a node and subtending a bud; it is usually expanded and concerned with the manufacture of food (photosynthesis).

Leaf bud – n. A bud which develops into a leafy branch; opposed to "flower bud."

Leaf, compound – n. A leaf with two or more blades, called leaflets.

Leaf cycle – n. In phyllotaxis, a spiral which passes through the insertion of intermediate leaves until it attains the next leaf exactly above its starting point.

Leafing – n. The unfolding of leaves.

Leaflet – n. A single division of a compound leaf.

Leaf scar – n. The mark or cicatrice left by the articulation and fall of a leaf.

Leaf sheath – n. The lower part of the petiole, which more or less invests the stem.

Leaf, simple – n. A leaf with a single blade; not compounded.

Leaf stock – n. The stem of a leaf; petiole, foot-stalk.

Leaf tendril – n. A tendril which is a transformed leaf.

Leaf trace – n. All the common bundles in a stem belonging to one leaf.

Lecotropal – a. Shaped like a horseshoe, as certain ovules.

Lectotype – n. A specimen or other element selected from the original material to serve as the nomenclatural type, when the holotype was not designated at the time of publication, or when the holotype is missing.

Legume – n. A member of the legume family, *Leguminosae*; a superior, one-celled monocarpellary fruit usually dehiscent into two valves and having the seeds attached along the ventral suture; indehiscent legumes are usually constricted between the seeds and break crosswise into one-seeded segments at maturity.

Leguminous – a. Pertaining to a legume, or to the *Leguminosae*.

Lemma (pl. **lemmata**) – n. In grasses, the flowering glume, the lower of the two bracts immediately enclosing the flower.

Lenitic – a. Pertaining to, or living in, quiet or still water.

Lenticel – n. A lenticular corky spot on young bark, corresponding to an epidermal stoma.

Lenticellate – a. Having lenticels.

Lenticular – a. Shaped like a double convex lens.

Lepanthium – n. A petal which contains a nectary.

Lepidoid – a. Scalelike; said of leaves, as *Thuja*.

Lepidophyllous – a. With scaly leaves.

Lepidopterid – n. A flower adapted to lepidopterous (butterfly and moth) pollination.

Lepis (pl. **lepides**) – n. A scale usually attached by its center.

Lepisma (pl. **lepismata**) – n. A membranous scale in some *Ranunculaceae,* an apparently aborted stamen in *Paeonia papaveracea.*

Leucanthous – a. White-flowered.

Liana – n. Luxuriant woody climbers in the tropics with anomalous structure.

Lid – n. Operculum.

Ligneous – a. Woody.

Lignified – a. Converted into wood.

Ligulate – a. With a ligule; strap-shaped or straplike.

Ligule – n. A strap-shaped body, such as the limb of the ray florets in *Compositae;* the lobe of the outer corona in *Stapelia*; the thin scarious projection from the top of the leaf-sheath in grasses; a narrow membranous, acuminate structure, internal to the leaf-base in *Isoëtes* and *Selaginella*; an appendage to certain petals as those of *Silene* and *Cuscuta*; the ovuliferous scale in *Araucaria*, united with the bract, and resembling the ligule in *Isoëtes*; the envelope which protects the young leaf in palms, as in *Chamacrops* and *Rhaphis*.

Liguliform – a. Strap-shaped.

Limb – n. The border or expanded part of a gamopetalous corolla, as distinct from the tube or throat; the lamina of a leaf or of a petal.

Limicolous – a. Growing in mud, as on the margins of pools, lakes, and ponds.

Limnodophilous – a. Dwelling in marshes.

Limnophilous – a. Dwelling in lakes.

Linear – a. Long and narrow with margins parallel, or nearly so.

Lineate – a. Lined; bearing thin parallel lines.

Lineolate – a. Marked with fine lines.

Lingulate – a. Tongue-shaped.

Linnean System – n. The artificial classification devised by Linnaeus, based upon the number and position of the stamens and pistils.

Linneon – n. A large species, usually polymorphic but with well characterized limits; species according to Linnaeus.

Lip – n. One of the two divisions of a bilabiate corolla or calyx, that is, a gamopetalous or gamosepalous organ cleft into an upper (superior or posterior) and a lower (inferior or anterior) portion; the labellum of orchids.

Littoral – a. Belonging to, or growing on the seashore.

Livid – a. Pale lead-colored.

Lobate – a. Divided into, or bearing lobes.

Lobe – n. Any part or segment of an organ; specifically, a part of petal, calyx, or leaf that represents a division to about the middle.

Lobulate – a. Divided into small lobes.

Lobule – n. A small lobe.

Lochmophilous – a. Dwelling in thickets.

Locule – n. Compartment or cell of a pistil or anther.

Loculicidal – a. With dehiscence on the back, between the partitions into the cavity.

Lodger-arrangement – n. The retention by certain flowers of their insect visitors.

Lodicule – n. A small scale outside the stamens in the flowers of grasses.

Loment – n. A flat legume which is constricted between the seeds, falling apart at the constrictions when mature into one-seeded joints.

Lomentaceous – a. Bearing or resembling loments.

Lorate – a. Strap-shaped, ligulate.

Lucid – a. Shining, referring to the surface.

Lunate – a. Crescent-shaped, half-moon shaped.

Lurid – a. Pale yellow.

Lustrous – a. Glossy, shiny.

Lutescent – a. Becoming yellow.

Luticolous – a. Growing in miry places.

Lyrate – a. Lyre-shaped, pinnatifid with the terminal lobe large and rounded, the lower lobes small.

M

Mace – n. The aril of the nutmeg.

Macrocladous – a. With long branches.

Macrophyllous – a. Having elongated or large leaflets or leaves.

Macrosporangium – n. The receptacle in which macrospores are produced.

Macrospore – n. The larger of the two kinds of spores in *Selaginella* and related plants.

Macrostylous – a. Long-styled.

Macula – n. A spot.

Maculate – a. Blotched or spotted.

Major – a. Larger.

Malacophilous – a. Said of plants that are pollinated by snails or slugs.

Malacophyllous – a. With soft leaves.

Malicorium – n. The rind of a pomegranate.

Malleolus – n. A layer; a shoot bent into the ground and half divided at the bend, whence it emits roots.

Malpighiaceous hairs – pl. n. Hairs which are straight and appressed but attached by the middle, frequent in *Malpighiaceae.*

Malvaceous – a. Malva-like, mallow-like.

Mammiform – a. Breast-shaped, conical with rounded apex.

Mammilla – n. A nipple or teat.

Mammillate – a. With little teat-shaped processes.

Mammose – a. With teat-shaped processes.

Manicate – a. Covered with a pubescence so thick and interwoven that it can be striped off like a sleeve.

Marbled – a. Stained with irregular streaks of color.

Marcescent – a. Withering without falling off.

Marginal – a. Of, pertaining to, or attached to the edge.

Marginate – a. Broad-brimmed, furnished with a margin of distinct character.

Marine – a. Growing within the influence of the sea, or immersed in its waters.

Maritime – a. Pertaining to the sea.

Massula – n. A group of cohering pollen-grains produced by one primary mother cell, as in orchids; also styled pollen-mass.

Mast – n. The fruit of such trees as beech, and other *Cupuliferae.*

Mattulla – n. The fibrous material surrounding the petioles of palms.

Matutinal – a. Pertaining to morning; plants flowering early, as *Impomoea purpurea.*

Median – a. Pertaining to the middle.

Megaphyllous – a. With large leaves or leaf-like expansions.

Megasporangium – n. The sporangium which produces the megaspores.

Megaspore – n. The more correct form of macrospore; the larger spore of heterosporous plants.

Megasporocarp – n. The product of the development of the megasporangium in *Azolla,* finally containing a single perfect megaspore.

Melangeophilous - a. Dwelling in loam.

Melanophyllous - a. Having leaves of a dark color.

Melanoxylon - n. Black wood.

Membranaceous - a. See Membranous.

Membranous - a. Thin, more or less flexible, and translucent; like a thin membrane.

Meniscoidal - a. Thin and concave-convex, like the crystal of a watch.

Mentum - n. An extension of the foot of the column in some orchids, in the shape of a projection in front of the flowers.

Mericarp - n. One of the two seedlike carpels of an umbelliferous fruit.

Meristem - n. Embryonic or undifferentiated tissue the cells of which are capable of active division.

Meristematic - a. Pertaining to the meristem.

Merotype - n. A specimen collected from the original type in cultivation by means of vegetative reproduction (Swingle).

Mesocarp - n. The middle layer of a pericarp.

Mesocotyl - n. An interpolated node in the seedling of some grasses, so that the sheath and cotyledon are separated by it.

Mesophyte - n. A plant intermediate between hydrophytes and xerophytes; a plant of medium moisture requirement.

Mesophytic - a. Growing under medium moisture conditions.

Metandry - n. A condition in which the female flowers mature before the male; protogyny.

Metatype - n. A specimen from the original locality, recognized as authentic by the describer himself.

Metoecious - a. Existing on different hosts; heteroecious.

Metonym - n. A synonymous name rendered invalid by the existence of an earlier valid name for the same species or other plant group.

Micropyle - n. The aperture in the skin of a seed formerly the foramen of the ovule; it marks the position of the radicle.

Microsorus - n. The male sorus in *Azolla.*

Microsporangium - n. The receptacle in which the microspores develop.

Microspore - n. The smaller of the two kinds of spores in such pteridophytes as *Selaginella.*

Midrib - n. The main rib or central vein of a leaf or leaflike structure.

Migration - n. Any movement by which the range of a species is extended. (Strictly speaking, it means moving under its own power.)

Migrule - n. The unit of migration, as seed, fruit, runner, bulb, etc.

Miniate - a. The color of red lead; more orange and duller than vermillion.

Minor - a. Smaller.

Minute - a. Very small, inconspicuous.

Misogamy - n. Reproductive isolation.

Mitriform - a. Shaped like a mitra or cap.

Mixed-inflorescence - n. One in which partial inflorescences develop differently from the main axis, as centrifugal and centripetal together.

Molendinaceous - a. Furnished with large, winglike expansions.

Monadelphous - a. With stamens united by their filaments into a tube or column.

Monandrous - a. With one stamen.

Monanthous - a. With one flower.

Moniliform – a. Necklace-shaped; like a string of beads.

Monocephalous – a. Bearing a single head or capitulum.

Monochasium – n. A one-branched cyme, either pure or resulting from the reduction of cymes.

Monochlamydeae – pl. n. A large division of phanerogams which have only one set of floral envelopes.

Monochlamydeous – a. Having only one set of floral envelopes.

Monoclinous – a. Having both stamens and pistils in the same flower; said of the capitula of composites which have only hermaphrodite flowers.

Monocotyledon – n. A plant having but one cotyledon or seed-leaf.

Monocotyledoneae – pl. n. Plants of the class identified by the possession of only one cotyledon.

Monocotyledonous – a. With one cotyledon or seed-leaf.

Monocyclic – a. With the members of a floral series in only one whorl, as with the calyx only.

Monodynamous – a. With one stamen much longer than the others.

Monoecious – a. Having unisexual flowers with both sexes borne on the same plant.

Monogynous – a. With one pistil.

Monolocular – a. One-celled, unilocular, applied to ovaries.

Monopetalous – a. One-petaled; gamopetalous, with the corolla composed of several petals laterally united.

Monophyllous – a. One-leaved, as an involucrum of a single piece; said of a leaf-bud in which a single leaf is subtended by an investing stipule; gamosepalous or gamopetalous.

Monopterous – a. One-winged.

Monosepalous – a. With one sepal.

Monospermous– a. One-seeded.

Monostachous – a. Arranged in one spike.

Monostichous – a. In a single vertical row.

Monostylous – a. Having a single style.

Monosymmetrical – a. Said of a flower which can be bisected equally in one plane only; zygomorphic; bilaterally symmetrical.

Monotrichous – a. Having one bristle.

Monotrophic – a. With nutrition confined to one host-species.

Monotropic – a. Said of bees which visit only one species of flower.

Monotype – n. A genus that contains only one species; the term is applicable to other categories.

Montane – a. Pertaining to mountains, as a plant which grows on them.

Moschate – a. Musky or musk-scented.

Moth-flowers – pl. n. Flowers adapted for moths as pollinating visitors; they are usually white, night-blooming flowers.

Motion-dichogamy – n. A condition in which the sexual organs vary in length or position during flowering.

Mucilaginous – a. Slimy, composed of mucilage.

Mucro – n. A sharp terminal point.

Mucronate – a. Furnished with a mucro (bristle-tipped).

Multiciliate – a. With many cilia.

Multicipital – a. With many heads, referring to the crown of a single root or to several caudices.

Multicostate – a. Many-ribbed, as the ribs running from the base of a leaf towards its apex.

Multifarious – a. Many-ranked, as leaves in vertical ranks.

Multifid - a. Cleft into many lobes or segments.
Multifoliate - a. Many-leaved.
Multipartite - a. Divided or cut many times.
Multiple fruit - n. A cluster of ripened ovaries traceable to the pistils of separate flowers, as the mulberry and the pineapple.
Multiplicate - a. Folded often or repeatedly.
Multiradiate - a. With numerous rays.
Multiseptate - a. With many partitions.
Muricate - a. Rough with short, hard points.
Muriculate - a. Very finely muricate.
Muriform - a. With brick-like markings, pits, or reticulations, as on some seed-coats.
Muscarian - a. With flowers that attract flies by a putrid stench.
Mutable - a. Able to produce mutants.
Mutant - a. That which undergoes mutation.
Mutation - n. A sudden hereditary variation of an offspring from its parents.
Muticous - a. Pointless, blunt, awnless, curtailed.
Myochrous - a. Mouse-colored.
Myrmecochorous - a. Dispersed by means of ants.
Myrmecochory - n. The state of being dispersed by ants.
Myremecodomate - n. A plant which provides shelters in which ants live.
Myrmecophilous - a. Said of plants which are inhabited by ants and offer specialized shelters or food for them; pollinated by ants.
Myrmecophobous - a. Shunning ants, said of plants which by hairs or glands repel ants.
Myrmecosymbiosis - n. The mutual relation between the ants and their host plant.

N

Nacreous - a. With pearly luster.
Naked - a. Wanting its usual covering, as without pubescence, or flowers destitute of perianth, or buds without scales.
Naked bud - n. A bud without scales.
Namated - n. A brook plant.
Napaceous - a. Turnip-shaped or rooted.
Napiform - a. Turnip-shaped.
Nascent - a. In the act of being formed.
Natant - a. Floating under water, that is, wholly immersed.
Naturalized - a. Having become thoroughly established in a region to which it is not indigenous.
Natural selection - n. The natural processes contributing to the "survival of the fittest".
Natural System - n. An arrangement according to the affinity of plants, and the sum of their characters, opposed to any artificial system, based on one set of characters.
Naucum - n. The fleshy part of a drupe; seeds with a very large hilum.
Naucus - n. Certain cruciferous fruits which have no valves.
Nautiloid - a. Spiral-formed like the shell of *Nautilus*.
Navicular - a. Boat-shaped; like the bow of a canoe.
Nebulose - a. Cloudy, misty, said of such finely divided inflorescence as of *Eragrostis*; smoke-colored.
Neck - n. The collar or junction of stem and root; the point where the blade separates from the sheath in certain leaves; the contracted part of the corolla or calyx tube.
Necrocoleopterophilous - a. Pollinated by carrion beetles.

Necrotype – n. A form that formerly existed, but is now extinct.

Nectar – n. The sweet secretion from glands or nectaries, which act as an inducement to insect visitors.

Nectar glands – n. The secreting organs which produce nectar.

Nectar guides – n. Lines, spots, or other devices directing to the nectary.

Nectariferous – a. Nectar-bearing.

Nectarostigma (pl. **stigmata**) – n. Some mark or depression indicating the prescence of a nectariferous gland.

Nectarotheca – n. The portion of a flower which immediately surrounds a nectaripore.

Nectary – n. The organ in which nectar is secreted, formerly applied to any anomalous part of a flower, as its spurred petal.

Needle – n. A stiff linear leaf as in *Pinaceae*.

Neism – n. The origin of an organ on a given place, as the formation of roots on a cutting.

Nema (pl. **nemata**) – n. A filament, a thread.

Nemus (pl. **nemores**) – n. Wood.

Neotype – n. A specimen selected to serve as the nomenclatural type of a taxon in a situation when all material on which the taxon was based is missing.

Nephroid – a. Reniform, kidney-shaped.

Nepionic – a. Said of the first leaves of seedlings developed immediately succeeding the embryonic stage of the cotyledons.

Nervation – n. Ventation, the manner in which the foliar nerves or veins are arranged.

Nerve – n. In botany, a simple or unbranched vein or slender rib.

Nervose – a. Full of nerves or prominently nerved.

Netted – a. Reticulated, net-veined with any system of irregularly anastomosing veins.

Neuter – a. Sexless, as a flower which has neither stamens nor pistils.

Nidulent – a. Partially encased or lying free in a cavity; embedded in a pulp, as the seeds in a berry.

Nigrescent – a. Turning black.

Nitid – a. Smooth and clear, lustrous, glittering.

Niveous – a. Snowy-white.

Nocturnal – a. Occurring at night, as night-blooming; active in the night.

Nodal – a. Relating to the node.

Nodal diaphragm – n. Any septum which extends across the hollow of a stem at the node.

Nodding – a. Curved somewhat from the vertical.

Node – n. That point on a stem which normally bears a leaf or leaves.

Nodiferous – a. Bearing nodes.

Nodose – a. Knotty or knobby.

Nodule – n. A small knot or rounded body.

Nodulose – a. With little knobs or knots.

Nomad – n. A pasture plant.

Nomenclature – n. The names of things in any science; in botany, frequently restricted to the correct usage of scientific names in taxonomy.

Nomen conservandum – n. A name retained in biological nomenclature regardless of priority.

Nomen novum (**nom. nov.**) – n. New name, i.e. a name hitherto unpublished, substituted for one in general use but found to be untenable.

Nomen nudum (**nom. nud.**) – n. A naked name, i.e. a name only; a plant name published without any description or figure, and hence which cannot be tied in with assurance to any plant or plant group. Nomina nuda are very properly rejected by all codes.

Nomogenesis - n. The theory that the evolution of organisms is the result of certain processes inherent in them and that it follows definite laws.

Nomophilous - a. Dwelling in pastures.

Normal - a. According to rule, usual as to structure.

Notate - a. Marked with spots or lines.

Nototribal - a. With the stamens and styles turned so as to strike their insect visitors on the back.

Nox - n. Night.

Nucamentaceous - a. Having the hardness of a nut; synonym for indehiscent, monospermal fruit; also, catkin-like.

Nucamentum - n. An ament or catkin.

Nuciferous - a. Nut-bearing.

Nucleus - n. A kernel of an ovule, which by fertilization becomes a seed; a dense protoplasmic structure near the center of living cells.

Nude - a. Bare, naked, uncovered.

Nut - n. A dry indehiscent, usually one-celled, one-seeded fruit (though usually traceable to a compound ovary) with a bony, woody, leathery, or papery wall and in general, partially or wholly encased in an involucre or husk.

Nutant - a. Nodding, drooping.

Nutlet - n. The diminutive of nut.

Nux - n. Nut.

Nyctanthous - a. Night-flowering.

Nyctigamous - a. Said of flowers which close by day, but open at night.

Nyctitropic - a. Turning in response to darkness.

Nyctitropism - n. The act of responding to darkness.

O

Obclavate - a. Club-shaped and attached at the thicker end.

Obcompressed - a. Compressed dorso-ventrally instead of laterally.

Obconic - a. Conical, but attached at the narrower end.

Obconical - a. Inversely conical, having the attachment at the apex.

Obcordate - a. Inversely heart-shaped, the notch being apical.

Obex (pl. **obices**) - n. A barrier; a hindrance to plant distribution. Biological obices, as the constitution of the plant. Physical obices, as the shutting in by mountains, etc.

Oblanceolate - a. Inverted lanceolate.

Oblate - a. Flattened at the poles, as a tangerine-orange.

Obligate - a. Necessary, essential, the reverse of facultative.

Oblique - a. Slanting, unequal-sided.

Oblong - a. Longer than broad, with the margins nearly parallel.

Obovate - a. Reversed ovate, the distal end the broader.

Obovoid - a. Appearing as an inverted egg.

Obscure - a. Dark or dingy in tint; uncertain in affinity or distinctiveness.

Obsolescent - a. Becoming rudimentary or extinct.

Obsolete - a. Not evident or apparent; rudimentary; no longer used.

Obturator - n. A small body accompanying the pollen mass of orchids and asclepiads closing the opening to the anther.

Obtuse - a. Blunt or rounded at the end.

Occlusion - n. The process by which wounds in trees are healed by the growth of callus.

Ocellus – n. A small eye.
Ochraceous – a. Ochre-colored.
Ochroleucous – a. Yellowish-white, buff.
Ochthad – n. A bank plant.
Ochthophilous – a. Bank-loving.
Ocrea – n. A legging-shaped or tubular structure formed by the union of two stipules.
Ocreate – a. With sheathing stipules.
Ocreola (pl. **ocreolae**) – n. The smaller or secondary sheaths, as in the inflorescences of *Polygonum*.
Octagynia – n. A Linnean order of plants with eight-styled flowers.
Octandrous – a. Having eight stamens.
Octolocular – a. Said of an eight-celled fruit or pericarp.
Octopetalous – a. With eight petals.
Octoradiate – a. With eight rays, as in some *Compositae*.
Octosepalous – a. With eight sepals.
Octostemonous – a. With eight fertile stamens.
Octostichous – a. In eight rows.
Oculus – n. The first appearance of a bud, especially on a tuber, as the eyes on a potato.
Odd pinnate – a. With a terminal leaflet, imparipinnate.
Official – a. Of the shops; used in medicine or arts.
Oleaginous – a. Oily and succulent.
Oleiferous – a. Oil-bearing.
Oleraceous – a. Having the qualities of garden herbs used in cooking.
Oligandrous – a. With few stamens.
Oligocarpic – a. Few-fruited.
Oligophyllous – a. Having few leaves.
Oligospermous – a. Few-seeded.
Oligotrophic – a. Growing in poor soil and competing for the nutritive salts in it.
Oligotropic – a. Said of insects which visit only a few species of plants.
Olivaceous – a. Olive green; olive-colored.
Ombrophilous – a. Rain-loving.
Ombrotropism – n. Tropic responses of organisms to the stimulus of rain.
Oncospore – n. A plant having seeds with hooks which aid in dispersion.
Onomatology – n. The science of names.
Ontogeny, ontogenesis – n. The developmental history of an individual from fertilized egg to adult organism.
Oöspore – n. The fertilized egg in the archegonium of cryptogams, from which the new plant develops directly.
Opaque – a. Dull; neither shining nor translucent.
Operculate – a. Furnished with a lid.
Operculum (pl. **opercula**) – n. A lid or cover which separates by a transverse line of division, as in the pyxis, also in some pollen grains.
Opposite – a. On both sides at the same level, as two leaves at a node; one part before another, as a stamen in front of a petal.
Oppositiflorous – a. Having opposite peduncles or pedicels.
Oppositifolious – a. With opposite leaves.
Optimal – a. The most advantageous for an organism or function.
Orange – a. The fruit of *Citrus aurantium*; a secondary color, red and yellow combined, taking its name from the tint of the fruit mentioned.
Orbicular – a. Flat with a circular outline.
Orbiculate – a. Disk-shaped, round.
Orchioid – a. Orchid-like.
Order – n. In botany, a group between genus (tribe and suborder) and class.

Orgadophilous – a. Dwelling in open woodland.

Organ – n. A group of tissues organized to perform a definite function.

Orifice – n. An opening by which spores, etc. escape; any opening.

Ornithogamous – a. Pollinated by birds.

Ornithophilous – a. Pollinated by birds.

Orophilous – a. Dwelling in mountainous regions.

Orthocladous – a. Straight-branched.

Orthogenesis – n. Purposive, "predetermined" evolution toward a definite objective; a tendency to vary continously in the same direction.

Orthoösmotropism – n. The assuming of an erect position to osmotic action.

Orthopterous – a. Straight-winged.

Orthostichous – a. Straight-ranked.

Orthotropism – n. The assumption of a vertical position.

Orthotropous – a. Said of an ovule or seed with a straight axis, chalaza at the insertion, the orifice at the other end.

Osseous – a. Bony.

Ossiculus – n. The pyrene or stone of a drupe.

Ossified – a. Becoming hard as bone, as the stones of drupes, such as the peach or plum.

Oval – a. Broadly elliptical with the width greater than half the length.

Ovary – n. That part of the pistil which contains the ovules.

Ovate – a. Shaped like a longitudinal section of a hen's egg, the broader end basal; applied to ovoid.

Ovoid – n. A solid that is oval (less correctly, ovate) in flat outline.

Ovulate – a. Pertaining to the ovule, or possessing ovules.

Ovule – n. That which becomes a seed after fertilization.

Ovuliferous – a. Bearing ovules.

P

Pabular – a. Of fodder or pasturage.

Pachycladous – a. Thick-branched.

Pagina – n. The blade or surface of a leaf.

Painted – a. Having colored streaks of unequal density.

Palate – n. In personate corollas, a rounded projection or prominence of the lower lip, closing the throat or very nearly so.

Pale – n. A chaffy scale such as often subtends the fruit of *Compositae*.

Palea – n. The inner bract of a grass floret; the chaffy scales on the receptacle of many *Compositae*; the ramenta or chaffy scales on the stipe of many ferns.

Paleaceous – a. Chaffy, chafflike in structure.

Paleobotany – n. Fossil botany, the study of plants in a fossil state.

Paleola (pl. **paleolae**) – n. A diminutive of palea, or of secondary order, as applied to the lodicule of grasses.

Paleolate – a. With a lodicule.

Paleophytological – a. Relating to the study of fossil plants.

Palet – n. See Palea.

Pallescent – a. Becoming light in tint.

Pallid – a. Pale.

Palmate – a. Resembling a hand with the fingers spread; having lobes radiating from a common point.

Palmately – a. In a palmate manner.

Palmatifid – a. Cut in a palmate fashion nearly to the petiole.

Paludose - a. Growing in marshy places.
Palustrine - a. Of or growing in marshes.
Pampiniform - a. Resembling the tendril of a vine.
Pampinus - n. Tendril.
Pandura - n. Violin.
Pandurate - a. Fiddle-like.
Panduriform - a. Fiddle-shaped.
Panicle - n. A compound or branched raceme.
Paniculate - a. Having a panicle type of inflorescence.
Panmixy - n. Free and more or less unlimited cross-ferilization.
Panniform - a. Having the appearance or texture of felt or woolen cloth.
Pannose - a. Having the appearance or texture of felt or woolen cloth of very close texture.
Papaveraceous - a. Belonging to or resembling the poppy.
Papery - a. Having the texture of paper, chartaceous.
Papilionaceous - a. Descriptive of the flower of many legumes having a standard, wings, and keel; with a pealike flower; like a butterfly.
Papilla (pl. **papillae**) - n. A minute nipple-shaped projection.
Papillary - a. Resembling papillae.
Papillose - a. Bearing papillae.
Pappiferous - a. Bearing pappus.
Pappus - n. Thistledown; the various tufts of hairs on achenes or fruits; the limb of the calyx of *Compositae* florets.
Papyraceous - a. Papery; white as paper.
Paracarpium - n. An abortive pistil or carpel; the persistent portion of some styles or stigmas.
Paracarpous - a. Said of ovaries whose carpels are joined together by their margins only.
Parachute - n. Sometimes applied to a fruit which is readily carried by wind, by means of membranous expansions or pappus, recalling the action of a parachute.
Paraheliotropism - n. Diurnal sleep, the movement of leaves to avoid the effect of intense sunlight.
Parallel - a. Extended in the same direction, but equally distant at every point.
Parallelodromous - a. Having parallel veins, as in lilies.
Parallel-veined - a. With lateral veins straight, as in *Alnus;* the entire system straight, as in the grasses.
Paraphototropism - n. The assumption of a position at right angles to the incident light.
Parasite - n. An organism subsisting on another, the host.
Parasitic - a. Deriving nourishment from another organism.
Parasol - n. A peculiar set of spines on some cacti.
Parastas (pl. **parastades**) - n. Used in the plural to designate the coronal rays of *Passiflora.*
Parastomon - n. An abortive stamen, a staminodium.
Paratype - n. A specimen belonging to the original series, but not the type selected by the author.
Paravariation - n. A modification or acquired variation developed during the life of the individual as a result of environmental causes and not heritable.
Parenchyma (pl. **parenchymata**) - n. Soft tissue of cells with unthickened walls, as pith cells.
Parietal - a. Borne on or belonging to a wall, as parietal placentation.
Paripinnate - a. Pinnate, with an equal number of leaflets, that is without a terminal one, abruptly pinnate.
Parted - a. Divided by sinuses which extend nearly to the midrib.

Parthenocarpy – n. The production of fruit without true fertilization.
Parthenogenesis – n. A form of apogamy in which the oösphere develops into a normal product of fertilization without a preceding sexual act.
Parthenogenetic – a. Developing without fertilization.
Partition – n. A wall or dissepiment; a separated part or segment; the deepest division into which a leaf can be cut without becoming compound.
Pastoral – a. Pertaining to shepherds; rural.
Patelliform – a. Disk-shaped.
Patent – a. Spreading.
Pathological – a. Diseased.
Patulous – a. Standing open, spreading.
Pectinate – a. Comblike; beset with narrow, closely inserted segments like the teeth in a comb.
Pedate – a. Palmately divided or parted with the lateral divisions cleft.
Pedatifid – a. Divided in a pedate manner nearly to the base.
Pedicel – n. An ultimate flower-stalk, the support of a single flower; in grasses, the stalk of a spikelet.
Pedicellate – a. Borne on a pedicel.
Pediculus – n. Pedicel; the stalk of an apple or other fruit; the filament of an anther.
Pediophilous – a. Dwelling in uplands or level country.
Peduncle – n. A primary flower stalk supporting either a cluster or a solitary flower.
Pedunculate – a. With a footstalk or peduncle.
Peg – n. An embryonic organ at the lower end of the hypocotyl of seedlings of *Cucumis, Gnetum,* etc., lasting until the cotyledons are withdrawn from the testa.
Pellicle – n. A delicate superficial membrane, epidermis.
Pellucid – a. Wholly or partially transparent.
Peloria, pelory – n. Reversion, on the part of the individual, to the production of regular flowers, when the species typically has asymmetrical or bilaterally symmetrical flowers.
Pelta – n. A bract attached by its middle as in peppers.
Peltafid – a. A peltate leaf cut into segments.
Peltate – a. Shield-shaped, as a leaf attached by its lower surface to a stalk instead of by its margin.
Pencilled – a. Marked with fine distinct lines.
Pendent – a. Hanging down from its support.
Pendulous – a. Hanging downward.
Penicillate – a. Like a pencil.
Pennate – a. Pinnate.
Pennatifid – a. Pinnatifid.
Penniveined – a. Veined in a pinnate manner.
Pentacamerous – a. With five locules.
Pentacarpellary – a. Having five carpels.
Pentacyclic – a. With five whorls of members.
Pentadactylous – a. Five fingered or with five finger-like divisions.
Pentadelphous – a. With five fraternities or bundles of stamens.
Pentagynous – a. With five pistils or styles.
Pentamerous – a. With parts in fives, as a corolla of five petals.
Pentandrous – a. With five stamens.
Pentapetalous – a. With five petals.
Pentapterous – a. Five-winged.
Pentasepalous – a. With five sepals.
Pentastichous – a. In five vertical ranks.
Pepo – n. The fruit of the gourd family, *Cucurbitaceae*; an inferior berry-like fruit with more or

less rind and with lateral placentation.

Perennation – n. A lasting, or perennial state.

Perennial – a. Continuing to live from year to year.

Perfect – a. Said of flowers having both sex organs present and functioning.

Perfoliate – a. Having the stem apparently passing through the leaf; said of opposite leaves joined at their bases.

Perforated – a. With holes.

Perianth – n. The floral envelope, of whatever form; the calyx and corolla.

Pericarp – n. The wall of a mature ovary.

Periclinium – n. The involucre of the capitulum in *Compositae*.

Peridroma – n. The rachis of ferns.

Perigynium – n. The hypogynous setae of sedges; the flask, or utricle of *Carex*; any hypogynous disc.

Perigynous – a. Borne around the ovary, as with calyx, corolla, and stamens borne on the edge of a cup-shaped hypanthium; such cases are said to exhibit perigyny. See also Hypogyny and Epigyny.

Peripheral – a. On or near the margin.

Peripterous – a. Surrounded by a wing or border.

Perisperm – n. The ordinary albumen of a seed, restricted to that which is formed outside the embryo sac; the pericarp or even the integuments of the seed.

Persicicolor – a. Peach-colored, rose pink.

Persistent – a. Remaining attached; not falling off.

Personate – a. Said of a bilabiate corolla having a prominent palate.

Perspicous – a. Transparent.

Perula – n. The scale of a leaf-bud; a projection in the flower of orchids, the mentum.

Perulate – a. Scale-bearing, as most buds.

Petal – n. One of the leafy expansions in the floral whorl styled the corolla; of the hop, the scales of the strobile.

Petaliferous – a. Bearing petals.

Petalode – n. An organ simulating a petal.

Petaloid – a. Like a petal, or having a floral envelope resembling petals.

Petiolar – a. Borne on, or pertaining to a petiole.

Petiolate – a. Having a petiole.

Petiole – n. The stem of a leaf.

Petiolule – n. A small petiole; the petiole of a leaflet.

Petricolous – a. Rock-inhabiting.

Petrophilous – a. Preferring rock.

Phaenantherous – a. With stamens exserted.

Phaenocarpous – a. Having a distinct fruit, with no adhesion to surrounding parts.

Phaenogamous – a. Plants sexually propagating by flowers, of which essential organs are stamens and pistils.

Phanerogam – n. A plant with flowers in which stamens and pistils are distinctly developed.

Pharmacognosy – n. The knowledge of the distinctive features of drugs.

Phellem – n. Cork.

Phellophilous – a. Dwelling in rock fields.

Phenological isolation – n. Isolation by time of flowering as either earlier or later than the other species of the genus.

Phenology – n. The science of the relations between climate and periodic biological phenomena, as the flowering and fruiting of plants, the migration of birds, etc.

Phenotype – n. A group of individuals of similar appearance but not necessarily of similar genetic constitution.

Phoeniceous – a. Purple-red.

Phoranthium – n. The receptacle of the capitulum of *Compositae.*

Photeolic – a. Pertaining to the "sleep" of plants.

Phototropism – n. The act of turning in response to light.

Phototype – n. A photograph of a type specimen, an abbreviation of the word photographotype.

Phragma (pl. **phragmata**) – n. A spurious dissepiment in fruits.

Phyllary – n. An involucral bract in the *Compositae.*

Phylloclad – n. A flattened branch assuming the form and function of foliage.

Phyllode – n. Leaflike petiole having no blade, as in some acacias and other plants.

Phylloid – a. Leaflike.

Phyllome – n. An assemblage of leaves or of incipient leaves in a bud.

Phyllopodes – pl. n. The dilated sheathing base of a leaf in *Isoëtes.*

Phylloptosis – n. The unnatural fall of leaves.

Phyllotaxy – n. The mode in which the leaves are arranged in regard to the axis.

Phylogenetic – a. Pertaining to the ancestral history of the race.

Phylogeny – n. The race history of an animal or plant deduced from development.

Phytogamy – n. Cross-fertilization of flowers.

Phytogenesis – n. The evolution and development of plants.

Phytogeographer – n. An expert on plant distribution.

Phytogeography – n. The science of plant distribution.

Phytography – n. The description and illustratiotn of plants; systematic or taxonomic botany.

Phytological – a. Relating to the study of plants.

Phytologist – n. A botanist.

Pileate – a. With a cap.

Pileorhiza – n. The root-cap, a hood at the extremity of the root.

Piliferous – a. Bearing hairs, or tipped with hairs; hair-pointed.

Piloglandulose – a. Bearing glandular hairs.

Pilose – a. With soft hairs.

Pinna – n. The primary unit of a pinnately compound leaf.

Pinnate – a. Feather-formed, as with the leaflets of a compound leaf placed on either side of a rachis.

Pinnately – adv. In a pinnate fashion.

Pinnately veined – n. With the vein pattern simulating a feather.

Pinnatifid – a. Cleft in a pinnate manner.

Pinnatisect – a. Cut down to the midrib in a pinnate way.

Pinninerved – a. Pinnately veined, running parallel towards the margin.

Pinnule – n. A secondary pinna; the foliaceous unit of a bipinnately compound leaf.

Pip – n. A popular name for the seed of an apple or pear.

Piperaceus – a. Peppery, pepper-like.

Piriform – a. Shaped like a pear.

Pisaceous – a. Pea-green, the color of unripe seeds.

Pisiferous – a. Pisum-bearing, pea-bearing.

Pisiform – a. Pea-shaped.

Pistil – n. The female organ of a flower, consisting when complete of an ovary, style, and stigma.

Pistillate – a. Having pistils and no stamens; female.

Pit – n. A small hollow or depression; the endocarp of a drupe containing a kernel or seed.
Pith – n. The spongy center of an exogenous stem.
Pitted – a. Marked with small depressions, punctate.
Place-constant – n. An invariable factor of plant life in a given locality.
Placenta – n. The place in the ovary where ovules are attached.
Placentation – n. The disposition of the placenta.
Plagiodromous – a. Said of tertiary leaf-veins when at right angles to the secondary veins.
Plagiophototropic – a. Assuming an oblique position to the rays of light, as leaflets of *Robinia*, *Tropaeolum*, etc.
Plaited – a. Plicate, folded like a fan.
Plane – n. Level, even, or flat surface.
Plane of symmetry – n. That which divides an object into symmetrical halves.
Plant – n. A vegetable organism nourished by gases or liquids and not ingesting solid particles of food.
Platanoid – a. Platanus-like, like the plane-tree or sycamore.
Pleiomery – n. The state of having more whorls than the normal number.
Pleiopetalous – a. Many-petaled.
Pleiopetaly – n. Doubleness in flowers.
Pleiosepalous – n. Many-sepaled.
Pleiospermous – a. With an unusually large number of seeds.
Pleurogyrate – a. Said of fern sporangia which have a horizontal annulus.
Pleurotribal – a. Said of flowers whose stamens are adapted to deposit their pollen upon the sides of insect visitors.
Plicate – a. Folded on the several ribs in the manner of a closed fan, occurring in palmately veined leaves, as in maple and currant.
Plococarpium – n. A fruit composed of follicles arranged around an axis.
Plotophyte – n. A floating plant, its functional stomata on the upper surface of its leaves.
Plumbeous–a. Lead-colored, greenish-drab.
Plumose – a. Pubescent in a manner simulating a feather or a plume.
Plumule – n. The primary leaf-bud of an embryo.
Plurilocular – a. Many-celled; with many locules.
Poad – n. A meadow plant.
Pod – n. A dehiscent dry pericarp; a rather general uncritical term.
Podocarp – n. A stipitate fruit, that is, one in which the ovary is borne on a gynophore.
Polachena – n. A fruit similar to a cremocarp, but composed of five carpels.
Pollard – n. A tree dwarfed by frequent cutting of its boughs a few feet from the ground, with subsequent thick growth of shoots from the place where cut.
Pollarding – n. The cutting back to produce a mop-headed growth.
Pollen – n. The fertilizing dust-like powder produced in the anthers of phanerogams, more or less globular in shape, sometimes spoken of as "microspores"; the male gametophyte in seed plants.
Pollen carrier – n. The retinaculum of asclepiads; the gland to which the pollen-masses are attached, either immediately or by caudicles.
Pollen flower – n. A flower which produces pollen but no nectar.

Pollen-mass – n. Pollen grains cohering by a waxy texture of fine threads into a single body.

Pollinate – v. To transfer pollen from the anther to the stigma or female organ.

Pollination – n. The placing of pollen on the stigma or stigmatic surface.

Polliniferous – a. Pollen-bearing.

Pollinium – n. A coherent mass of pollen, as in orchids and asclepiads.

Polster – n. A cushion plant, a low, compact perennial.

Polyadelphous – a. With stamens grouped into several brotherhoods or bundles.

Polyandrous – a. Having an indefinite number of stamens.

Polyanthous–a. Having many flowers, particularly if within the same involucre.

Polycephalous – a. Bearing many heads or capitula.

Polycotyledonous – a. With more than two cotyledons.

Polyembryony – n. The presence of more than one embryo in an ovule.

Polygamodioecious – a. Polygamous but chiefly dioecious.

Polygamomonoecious – a. Polygamous but chiefly monoecious.

Polygamous–a. Bearing perfect and unisexual flowers on the same individual.

Polygenesis – n. The production of a new type at more than one place or more than one time.

Polymerous – a. With numerous members to each series or cycle.

Polymorphic – a. With several or various forms, variable as to habit.

Polypetalous – a. With many distinct petals.

Polysepalous – a. With many distinct sepals.

Polystachyous – a. Having many spikes.

Polystemonous – a. Polyandrous; with numerous stamens.

Polystichous – a. With many ranks or rows, as leaves.

Polythalamic – a. Having more than one female flower within the involucre; derived from more than one flower, as a collective fruit.

Polytropism – n. The turning of leaves in order to place themselves vertically and meridionally, the two surfaces facing east and west.

Pomaceous – a. Relating to apples.

Pome – n. A fleshy fruit, the product of a compound pistil with the seeds encased within a papery or cartilaginous cell, as the apple.

Pomeridian – a. Afternoon, as blooming in the afternoon.

Pontohalicolous – a. Dwelling in salt marshes.

Porandrous – a. With anthers which open by pores.

Poricidal – a. Opening by pores.

Porose – a. With small holes or pores.

Porrect – a. Directed outward and forward.

Posterior – a. At or toward the back; opposite the front; toward the axis; away from the subtending bract.

Potamophilous – a. River-loving.

Praemorse – a. As though bitten off, terminated abruptly.

Precocious – a. Appearing or developing very early, as the aments in *Salix* expanding before the leaves.

Preformed – a. Said of flowers and inflorescences which appear in fall, but do not function until the following spring.

Prehensile – a. Clasping or grasping, as in tendrils.

Prevernal – a. Early spring-flowering; of early spring.

Prickle – n. A small and weak spinelike body borne irregularly on the bark or epidermis.

Primary – a. First in order of time or development.

Primine – n. The outer integument of an ovule.

Primocane – n. The first year's cane (usually without flowers) of *Rubus* and similar genera.

Prismatic – a. Of the shape of a prism.

Prison flowers – pl. n. Flowers which imprison their insect-visitors until fertilization is effected.

Proanthesis – n. A flowering in advance of the normal period, some flowers appearing in autumn in advance of the ensuing spring.

Process – n. Any projecting appendage.

Procumbent – a. Prostrate, trailing; lying flat upon the ground.

Prohydrotropism – n. The act of turning toward a source of moisture.

Proliferating – a. Producing offshoots.

Propagule – n. See Diaspore.

Prophyllum – n. The bracteole at the base of an individual flower, as in *Juncus*; a membranous structure between a branch and the main stem in *Graminae*.

Prostrate – a. Laying flat on the ground.

Protandrous – a. With anthers maturing before the pistils in the same flower.

Protandry – n. A condition in which the anthers mature before the pistil in the same flower, the pollen being dispersed before the pistil is receptive.

Protanthesis – n. The normal first flower of an inflorescence.

Proterandrous – a. With anthers ripening before the pistil in the same flower; protandrous, one kind of dichogamy.

Proterogyny – a. A condition in which the pistil is receptive before the anthers have mature pollen.

Proterotypes – pl.n. Primary types; all specimens which have served as the basis for descriptions and figures of organisms; further divided into Holotype, Cotype (or Syntype), Paratype, Lectotype, and Chirotype.

Prothallium – n. The minute reduced gametophyte of the ferns and their allies (*Pteridophyta*).

Prothallus – n. The gametophyte stage or generation of Pteridophytes, a multicellular and usually flattened thallus-like structure on the ground, bearing the sexual organs, as the antheridia and archegonia.

Protogenesis – n. Reproduction by budding.

Protogynous – a. Characterized by protogyny.

Protogyny – n. A condition in which the pistil matures before the anthers.

Protolog – n. The original description of a genus, species, or variety.

Prototype – n. The assumed ancestral form, from which the descendents have become modified.

Pruinose – a. With a waxy powdery secretion on the surface, glaucous.

Pruniform – a. Plum-shaped.

Prurient – a. Causing an itching sensation.

Psammophilous – a. Sand-loving, as the vegetation of dunes.

Pseudobulb – n. The thickened or bulb-formed stems of certain orchids, the part being solid and borne above ground.

Psilicolous – a. Prairie-dwelling.

Psilophilous – a. Prairie-loving.

Psychophilous – a. Pollinated by diurnal lepidoptera.
Psychrocleistogamy – n. Cleistogamy induced by cold.
Pterocarpous – a. Wing-fruited.
Pterocaulous – a. Wing-stemmed.
Pterospermous – a. With winged seeds.
Pterygopous – a. Having the peduncle winged.
Puberulent – a. Somewhat or minutely pubescent.
Pubescence – n. Hairiness.
Pubescent – a. Covered with short soft hairs, or down.
Pugioniform – a. Dagger-shaped.
Pulveraceous – a. Covered with a layer of powdery granules.
Pulverulent – a. Powdered, as if dusted over.
Pulvinate – a. Cushion-shaped.
Pulviniform – a. Having the shape of a cushion or pad.
Pulvinus – n. An enlargement close under the insertion of a leaf; the swollen base of a petiole, as in *Mimosa pudica,* sometimes at the top of the petiole.
Pumpform – a. Applied to papilionaceous flowers, with concealed anthers, as *Lotus, Coronilla,* and *Ononis.*
Punctate – a. Marked with dots, depressions or translucent glands.
Puncticulate – a. Minutely punctate.
Pungent – a. Ending in a rigid and sharp point.
Puniceous – a. Crimson, reddish-purple.
Purpurescent – a. Becoming or turning purple.
Pustular – a. Having slight blister-like elevations.
Pustulose – a. Blistery, furnished with pustules or irregular raised pimples (not as roughened as papillose.)
Putamen – n. The shell of a nut; the hardened endocarp of stone fruit.
Pyramidal – a. Pyramid-shaped.
Pyrene – n. Nutlet, particularly the nutlet in a drupe.
Pyriform – a. Pear-shaped.
Pyxidate – a. With a lid, as some capsules.
Pyxis – n. A capsule with circumscissile dehiscence, the upper portion acting as a lid.

Q

Quadrangulate – a. Having four angles, which are usually right angles.
Quadrate – a. Nearly square in form.
Quadrifoliate – a. With four leaves or leaflets.
Quilled – a. Said of normally ligulate florets which have become tubular.
Quinary – a. In fives.
Quinate – a. Growing together in fives, as leaflets from the same point.
Quincuncial – a. Arranged in a quincunx; in aestivation, partially imbricated of five parts, two being exterior, two interior, and the fifth one having one margin exterior and the other interior, as in the calyx of the rose.
Quinquecostate – a. Having five ribs.
Quinquefarious – a. In five ranks.
Quinquefoliate – a. Five-leaved.
Quinquejugate – a. In five pairs, as of leaflets.
Quinquelocular – a. Five-celled.
Quinquenerved – a. With the midrib dividing into five, that is, the main rib and a pair on each side.

R

Race – n. A variety of such fixity as to be reproduced from seeds; used also in a loose sense for related individuals without regard for rank.

Raceme – n. An indeterminate inflorescence consisting of a central rachis bearing a number of flowers with pedicels of nearly equal length.

Racemiform – a. In the form of a raceme.

Racemose – a. Resembling a raceme; in racemes.

Rachilla – n. A diminutive or secondary rachis or axis; in grasses and sedges, the axis that bears the florets.

Rachis – n. An axis bearing flowers or leaflets; petiole of a fern frond.

Radiant – a. Diverging from a central point.

Radiate – a. Spreading from or arranged around a common center.

Radical – a. Belonging or pertaining to the root.

Radicant – a. Rooting, usually applied to stems or leaves.

Radicicolous – a. With the flower seated immediately upon the crown of the root; dwelling in the root as a parasite.

Radicle – n. The lower portion of the axis of an embryo seedling.

Rain-leaves – pl. n. Leaves which are adapted to shed the rain from their surfaces, and generally are acuminate-tipped; drip tips.

Ramal – a. Belonging to a branch.

Rameal – a. See Ramal.

Ramentum (pl. **ramenta**) – n. Used in the plural for the thin chaffy scales of the epidermis, as the scales of many ferns.

Ramose – a. Branching, having many branches.

Ramulose – a. Having many branches.

Rank – n. A row, especially a vertical row.

Raphe – n. In a more or less anatropous ovule, a cord or ridge of fibro-vascular tissue connecting the base of the nucellus with the placenta.

Raphis (pl. **raphides**) – n. A needle-shaped crystal, used in the plural to describe the crystals found in the cells of some plants.

Rapiformis – a. Turnip-shaped.

Ratoon – n. A shoot from the root of a plant which has been cut down.

Ray – n. One of the radiating branches of an umbel; the marginal, as opposed to the disk, flowers in *Compositae* or other flower clusters, when there is a difference in structure.

Ray flower – n. An outer ligulate flower of *Compositae*.

Ray, medullary – n. The primary rays in exogenous stems between the different bundles, passing radially outwards.

Recapitulation theory – n. That every organism in its individual life-history repeats the various stages through which its ancestors have passed in the course of evolution.

Receptacle – n. That expanded portion of the axis which bears the floral organs; torus.

Reclinate – a. Bent down or falling back from the perpendicular.

Recurved – a. Bent or curved downward or backward.

Reduplicate – a. Doubled back; as a term of aestivation, in which the edges are valvate and reflexed.

Reflexed – a. Abruptly curved or bent downward or backward.

Regal – a. Royal.

Region – n. The area occupied by given forms.

Region, austral – n. Southern region.

Region, boreal – n. Northern region.

Region, tropical – n. Region within the tropics.

Regma (pl. **regmata**) – n. A fruit with elastically opening segments or cocci, as in *Euphorbia*; a form of schizocarp.

Regular – a. Uniform or symmetrical in shape or structure; of a flower, actinomorphic, radially symmetrical.

Relic, relict – n. A species properly belonging to an earlier vegetation type than that in which it is now found.

Remote – a. Scattered, not close together.

Reniform – a. Kidney-shaped, said of the form of some leaves.

Repand – a. Undulate or wavy, as the margin of some leaves.

Repent – a. Creeping, prostrate, and rooting at the nodes.

Repletum–n. A fruit with the valves connected by threads, persistent after dehiscence, such as in orchids, *Aristolochia* and some *Papaveraceae*.

Replum – n. A framelike placenta from which the valves fall away in dehiscence, frequently used so as to include the septum of *Cruciferae*.

Reprogression – n. A mode of flowering in which the primordial flower at the summit opens first, after which flowering occurs in succession from the bottom upwards.

Reptant – a. Creeping on the ground and rooting.

Resilient – a. Springing or bending back, as some stamens.

Resin cyst – n. Cell or cavity occluded with resin.

Resiniferous – a. Producing resin.

Rest – n. Dormancy induced in cold climates by lowness of temperature, in hot climates by want of moisture.

Resupinate – a. Upside down, or apparently so.

Reticulate – a. Forming a network.

Reticulation – n. Network, the regular crossing of threads.

Reticulum –n. A membrane of cross-fibers found in palms at the base of the petiole.

Retinaculum – n. The gland to which one or more pollinia are attached in orchids; in asclepiads, a horny elastic body to which the pollen-masses are fixed; in most *Acanthaceae*, the funiculus, which is curved like a hook and retains the seed until mature.

Retrocurved – a. Bent back.

Retroflexed – a. Bent back, reflexed.

Retrorse – a. Directed backward or downward.

Retuse – a. With a shallow notch at a rounded apex.

Reversion – n. A change backward, as to an earlier condition.

Revolute – a. Rolled backward, margin rolled toward lower side.

Rhabdocarpous – a. Long-fruited; with fruits shaped like a rod.

Rhachilla – n. See Rachilla.

Rhachis – n. See Rachis.

Rhaphe – n. See Raphe.

Rhaphis (pl. **rhaphides**) – n. See Raphis.

Rheophilous – a. Creek-loving.

Rheotropism – n. A turning in response to a current of water.

Rhipidium – n. A fan-shaped cyme, the lateral branches being developed alternately in two opposite directions.

Rhizanthous – a. Root-flowered; flowering from the root or seeming to do so.

Rhizocarp – n. A sporangium such as is produced on rootlike processes of members of the *Marsileacae*.

Rhizomatose – a. Having the character of a rhizome.

Rhizome – n. The rootstock or dorsiventral stem having a rootlike appearance, prostrate on or under ground, sending off rootlets, the apex progressively sending up stems or leaves.

Rhizophilous – a. Growing attached to roots.

Rhizophyllous – a. Roots that proceed from leaves.

Rhizotaxis – n. The system of arrangement of roots.

Rhomboidal – a. Approaching a rhombic outline, quadrangular, with the lateral angles obtuse.

Rhyacophilous – a. Torrent-loving.

Rib – n. A primary vein, especially the central longitudinal or midrib.

Rictus – n. The mouth or gorge of a bilabiate corolla.

Rigescent – a. Becoming rigid.

Rigid – a. Stiff, inflexible.

Rimose, rimous – a. With chinks or cracks, as in old bark.

Rind – n. The outer bark of a tree, all the tissue outside the cambium; the tough outer layer of some fleshy fruits.

Ringent – a. Gaping, as the mouth of an open bilabiate corolla.

Riparious – a. Growing by rivers or streams.

Ripe – a. Mature, characterized by the completion of an organ or organism for its allotted function.

Rivulose – a. Having small sinuate channels; marked with lines like a rivulet.

Rogue – n. A gardener's name for a plant which does not come true from seed; a variation from the type.

Root – n. The descending axis, growing opposite from the stem, without nodes, mostly developing underground, and absorbing moisture from the soil.

Rootlet – n. A very slender root or the branch of a root.

Rootstock – n. Subterranean stem; rhizome.

Rosette – n. A cluster of spreading or radiating basal leaves.

Rostellate – a. The diminutive of rostrate, somewhat beaked.

Rostellum – n. A little beak; a slender extension from the upper edge of the stigma in orchids.

Rostrate – a. With a beak, narrowed into a slender tip or point.

Rostrum – n. Any beak-like extension; the inner segment of the coronal lobes in asclepiads.

Rosula – n. Small rose; a rosette of leaves, as in houseleek.

Rosulate – a. In the form of a rosette.

Rotate – a. Wheel-shaped, circular, and flat, applied to a gamopetalous corolla with a short tube.

Rotund – a. Rounded in outline, somewhat orbicular, but a little inclined toward the oblong.

Rotundifolious – a. Round-leaved.

Rubescent – a. Reddish, becoming red.

Rubiginose – a. Rust-colored, usually applying to glandular hairs.

Ruderal – n. A plant growing in rubbish or waste places.

Rudiment – n. An imperfectly developed organ or part.

Rudimentary – a. Arrested in an early stage of development.

Rufescent – a. Becoming reddish.

Rufous – a. Reddish-brown.

Rugose – a. Wrinkled, as a leaf surface with sunken veins.

Rugulose – a. Finely wrinkled.

Ruminate – a. Having a chewed appearance.

Runcinate – a. Saw-toothed or sharply incised, the teeth retrorse.

Runner – n. A stolon, an elongated lateral shoot, rooting at intervals, the intermediate part apt to perish, and thus new individuals arise.

Rupestral – a. Growing on walls and rocks.

Rupicolous – a. Growing among the rocks.
Ruptile – a. Dehiscing in an irregular manner.
Rupturing – a. Bursting irregularly.

S

Sabulicolous – a. Growing in sandy places.
Saccate – a. Bag-shaped.
Sacciform – a. Bag-shaped.
Sacellus – n. A one-seeded indehiscent pericarp, enclosed within a hardened calyx, as the *Marvel of Peru;* applied to such fruits as those of *Chenopodium* which burst irregularly.
Sagittate – a. Enlarged at the base into two acute straight lobes, like the barbed head of an arrow.
Saline – a. Of or pertaining to salt.
Salverform – a. With a slender tube and an abruptly expanding limb, as that of the *Phlox*; hypocrateriform, salver-shaped.
Salver-shaped – a. See Salverform.
Samara – n. A winged achene-like fruit.
Samaroid – a. Samara-like.
Sanguine – a. Blood-colored, crimson.
Sap – n. The juice of a plant; the fluid contents of cells and young vessels, consisting of water and salts absorbed by the roots and distributed through the plant.
Sapid – a. Having a pleasant taste.
Saponaceous – a. Soapy, slippery to the touch.
Sapor – n. The property of the taste of a plant, such as bitterness.
Sapromyiophilous – a. Growing in humus.
Saprophyte – n. A plant deriving all of its nourishment from the bodies of decaying organisms.
Sap wood – n. The new wood in an exogenous tree, so long as it is pervious to the flow of water; the alburnum.
Sarcospore – n. A plant with fleshy disseminules.
Sarment – n. A long, slender runner, or stolon, as in the strawberry.
Sarmentose – a. With long slender runners.
Saurochore – n. A plant disseminated by lizards or snakes.
Sausage-shaped – a. Allantoid.
Sautellus – n. A bulblet such as those of *Lilium tigrinum.*
Saxicolous – a. Dwelling or growing among rocks.
Scabridulous – a. Slightly rough.
Scabrous – a. With short bristly hairs; rough to the touch.
Scalariform – a. Having markings suggestive of a ladder.
Scale – n. Any thin scarious body, usually a degenerative leaf, sometimes of epidermal origin.
Scandent – a. Climbing in any manner.
Scape – n. Leafless peduncle arising from the ground, it may bear scales or bracts but not foliage-leaves and may be one- or many-flowered.
Scaphoid – a. Boat-shaped.
Scapiflorous – a. Having flowers borne on a scape.
Scapiform – a. Resembling a scape.
Scapose – a. Bearing or resembling a scape.
Scar – n. A mark left on the stem by the separation of a leaf, or on a seed by its detachment; a cicatrix.
Scarious – a. Thin, dry, and membranous, not green.
Scarlet – a. Vivid red, having some yellow in its composition.
Schistaceous – a. Slate gray.
Schizocarp – n. A pericarp which splits into one-seeded portions, mericarps.
Schizopetalous – a. With cut petals.

Scion – n. A young shoot, a twig used for grafting.
Sciophilous – a. Shade-loving.
Sciophyll – n. A shade leaf.
Scissile – a. Separating, easily split.
Sciuroid – a. Curved and bushy like a squirrel's tail.
Scleranthium – n. An achene enclosed in an indurated portion of the calyx tube, as in *Mirabilis.*
Scleroid – a. Having a hard texture.
Sclerophyllous – a. Hard-leaved.
Scobiform – a. Having the appearance of sawdust.
Scobina – n. The zigzag rasplike rachilla of the spikelet of some grasses.
Scobinate – a. With a surface that feels rough as though rasped.
Scorpioid – a. Said of a coiled cluster in which the flowers are two-ranked and borne alternately at the right and left.
Scorpioid cyme – n. Cincinnus, the lateral branches developed on opposite sides alternately, as in *Boraginaceae.*
Scotophilous – a. Dwelling in darkness.
Scrobiculate – a. Marked by minute or shallow depressions, pitted.
Scrotiform – a. Pouch-shaped.
Scurf – n. Small branlike scales on the epidermis.
Scurfy – a. Covered with small branlike scales.
Scutate – a. Buckler-shaped, like a small shield.
Scutellate – a. Shaped like a small platter.
Scutellum –n. Any of several small shield-shaped parts or organs; a conical cap of the endosperm in *Cycadeae*; the first leaf in a grass embryo attached at the basal node of the mesocotyl and serving as a food storage organ.
Scutum – n. The broad dilated apex of the style in asclepiads.
Seasonal amphichromatism – n. The production of two differently colored flowers on the same stock due to season.
Seasonal heterochromatism–n. The production of different colors in the flowers of the same inflorescence due to season.
Sebaceous – a. Like lumps of tallow.
Secondary peduncle – n. A branch of a many-flowered inflorescence.
Secund – a. Said of parts or organs directed to one side only, usually by torsion.
Seed – n. A mature ovule.
Seed-leaf – n. Cotyledon.
Seedling – n. A plant produced from seed, in distinction to a plant propagated vegetatively; a juvenile plant.
Seed-stalk – n. The funiculus or podosperm.
Segment – n. One of the parts of a leaf, petal, calyx, or perianth that is divided but not truly compound; any of the parts into which an organism naturally separates or is divided; a section.
Segregate – n. That which is kept apart; a segregate is a species separated from a super-species.
Sejugous – a. Having six pairs of leaflets, as some pinnate leaves.
Selenotropism – n. Movements of plants caused by the light of the moon.
Selfed – a. Fertilized by its own pollen.
Self-fertilization – n. Fertilization by its own pollen.
Self-pollination – n. Pollination by pollen from the same flower.
Semen (pl. **semines**) – n. The seed of a flowering plant.
Semester ring – n. The ring produced in the wood of many trop-

ical trees, in consequence of two periods of growth and rest in a year.

Semiglobose - a. Half-globose; hemispherical.

Semilunate - a. Shaped like a half-moon, crescent-shaped.

Seminiferous - a. Seed-bearing; used for the special portion of the pericarp bearing the seeds.

Seminiferous scale - n. In *Coniferae*, that scale above the bract-scale on which the ovules are placed and the seeds borne.

Senescence - n. Aging of protoplasm.

Senescent - a. Growing old or effete.

Sensitive - a. Responsive to stimuli, as the leaves of *Mimosa pudica*.

Sepal - n. One of the separate parts of a calyx.

Sepaloid - a. With the texture of, or resembling a sepal.

Sepicolous - a. Inhabiting hedges.

Septate - a. Partitioned; divided by partitions.

Septentate - a. Having parts in sevens, as in a compound leaf, with seven leaflets arising from the same point.

Septicidal - a. Said of a capsule that dehisces through the dissepiments or lines of junction.

Septifolious - a. Seven-leaved.

Septifragal - a. With the valves breaking away from the dissepiments in dehiscence.

Septum - n. Any kind of partition, whether a true or false dissepiment.

Seriate - a. Disposed in series of rows, either transverse or longitudinal.

Sericeous - a. Silky, clothed with closely appressed, soft, straight pubescence.

Serotinal - a. Produced late in the season, or the year, as in autumn; autumnal.

Serotinous - a. Produced or occurring late in the season.

Serra - n. The tooth of a serrate leaf.

Serrate - a. With sharp teeth on the margin pointing forward.

Serrulate - a. Serrate with minute teeth.

Sessile - a. Without a stalk of any kind, as a leaf without a petiole.

Seta (pl. **setae**) - n. A bristle or bristle-shaped body; the arista or awn of grasses, when terminal; a peculiar stalked gland in *Rubus*; used by cyperologists for the bristle within the utricle of certain species of *Carex*; it represents a continuation of the floral axis.

Setaceous - a. Bristle-like, with bristle.

Setiferous - a. Bearing bristles.

Setiform - a. In the shape of a bristle.

Setose - a. Bristly, beset with bristles.

Setulose - a. With minute bristles.

Sex - n. In botany, male or female functions in plants.

Sexual - a. Pertaining to sex.

Shade leaves - n. Leaves adapted to modified light.

Sheath - n. Any long or more or less tubular structure surrounding an organ or part.

Sheathing - a. Enclosing, as by a sheath.

Shield - n. The staminode of *Cypripedium*; in *Coniferae*, the thick rhomboid extremity of the cone-scales.

Shield-shaped - a. In the form of a shield.

Shoot - n. A young growing branch or twig; the ascending axis.

Shrub - n. A low, usually several-stemmed, woody plant; a bush.

Sigmoid – a. Said of a leaflet or segment that is curved sidewise in opposing directions; S-shaped.
Siliceous, silicious – a. Composed of or abounding in silica.
Silicle – n. The short fruit of certain *Cruciferae,* silicule.
Silicolous – a. Growing in flinty soils.
Silicule – n. A short silique, not much longer than broad, silicle.
Silique – n. The peculiar pod of the *Cruciferae,* with two valves falling away from a frame (the replum) on which the seeds grow, and across which a false partition is formed.
Silks – pl. n. The styles and stigmas in maize.
Silky – a. Having a covering of soft appressed fine hairs; sericeous.
Silva – n. See Sylva.
Silvery – a. With a whitish metallic more or less shiny luster.
Simple – a. Of one piece; not compound.
Simple fruits–n. Those fruits which result from the ripening of a single pistil.
Simple inflorescence – n. A flower cluster with one axis, as a spadix, spike, or catkin.
Simple leaf – n. A leaf with one blade, with incomplete segmentation.
Simple pistil – n. A pistil consisting of one carpel.
Sinistrorse – a. Turning to the left; counterclockwise.
Sinuate – a. With a deep, wavy margin.
Sinuous – a. See Sinuate.
Sinus – n. The space between two lobes of a leaf or other expanded organ.
Siotropism – n. Response to shaking, as with *Mimosa.*
Skin – n. A thin external covering, the cuticle or epidermis.
Skotophilous – a. See Scotophilous.
Sleep – n. The response of plants to the absence of light resulting in changes in position of organs such as leaves.
Sleep movement – n. Positions taken by leaves during the night; nyctitropic movement.
Smooth – a. Without roughness or pubescence.
Snail-plants – pl. n. Plants which are supposed to be pollinated by snails and slugs; malacophilous plants.
Sobole – n. A shoot, especially from the ground.
Soboliferous – a. Sucker-bearing.
Sole – n. Of a carpel, the end farthest from the apex.
Soleaform – a. Slipper-shaped, almost resembling an hourglass.
Solitary – a. Single, only one from the same place; species of which the individuals are in extreme isolation.
Sordid – a. Dirty in tint, chiefly applied to pappuses of an impure white.
Sorophore – n. A gelatinous cushion of the ventral edge of the sporocarp of *Marsilea* and ferns.
Sorose – a. A fleshy multiple fruit, as a mulberry or pineapple.
Sorus – n. A cluster of sporangia in ferns.
Spadiceous – a. As to color, date brown; having the nature of, or bearing a spadix.
Spadix – n. The thick or fleshy spike of certain plants, as the *Araceae,* surrounded or subtended by a spathe.
Spananthus – a. Having new flowers.
Sparse – a. Scattered.
Spathe – n. The bract or pair of bracts surrounding or subtending a flower cluster or spadix; it is sometimes colored and flower-like, as in the *Calla.*

Spathella – n. An old name for the glumes of grasses; sometimes the paleae were included.

Spathellula – n. A palea of grass.

Spathe valves – n. The bractlike envelopes beneath the flowers in certain monocotyledons, as *Allium* and *Narcissus.*

Spatulate – a. Spatula-shaped.

Speciation – n. The processes whereby new species are formed.

Species (pl. **species**) – n. Groups of actually or potentially interbreeding natural populations, which are reproductively isolated from other such groups.

Species nova – n. New species.

Specific name – n. The latin appellative appropriated to a given species, usually an adjective, but sometimes a substantive used in apposition.

Specimen – n. A plant, or portion of one, prepared for botanic study.

Speiranthy – n. The state of having a twisted flower.

Spermatophyte – n. A phanerogam, a plant with true seeds.

Spermatozoid – n. A free swimming male gamete.

Sphalerocarpum – n. An accessory fruit, as an achene in a baccate calyx-tube.

Sphenoid – a. Wedge-shaped, cuneate.

Sphingophilous – a. With flowers pollinated by hawkmoths and nocturnal lepidoptera; they usually have a strong sweet smell, and nectar in flower-tubes.

Spicate – a. Like a spike, or disposed in a spike.

Spiciform – a. Spikelike, in the form of a spike.

Spicule – n. A diminutive or secondary spike; a fine, fleshy erect point.

Spiculose – a. With a surface covered with fine points.

Spike – n. An inflorescence consisting of a central rachis bearing a number of sessile flowers.

Spikelet – n. The unit of the inflorescence in grasses, consisting of two glumes and one or more florets; a diminutive spike.

Spikelike – a. Said of a dense panicle in which the pedicels and branches are short and hidden by the spikelets, as in *Phleum.*

Spiladophilous – a. Dwelling in clay.

Spindle-shaped – a. Fusiform, tapering from the middle toward each end.

Spine – n. A sharp woody outgrowth from the stem, usually a modified branch, sometimes a petiole, stipule, or other part.

Spinescent – a. Ending in a spine or sharp point; more or less spiny.

Spinose – a. Spinelike or with spines.

Spinule – n. A little spine or spinelike process.

Spinulose – a. With small spines.

Spiny – a. Beset with spines.

Spiral – a. As though wound around an axis.

Spiral flower – n. A flower with the members arranged in spirals and not in whorls.

Spiricle–n. A delicate coiled thread in the superficial cells of certain seeds and achenes which uncoils when moistened, as in *Ruellia.*

Splint – n. A forester's term for alburnum or sapwood.

Sponsalia – n. Anthesis; the pollination period.

Sporadic – a. Occurring here and there, without continuous range.

Sporangiophore – n. A stalklike structure bearing sporangia.

Sporangium–n. A sac endogenously producing spores.

Spore – n. A cell which becomes free and capable of direct development into a new individual.

Sporocarp – n. A receptacle containing sporangia or spores.
Sporophore – n. A spore-bearing branch or organ.
Sporophyll – n. A spore-bearing leaf.
Sport – n. A sudden spontaneous deviation or variation of an organism from type, beyond the usual limits of individual variation.
Spray – n. Small branches or branchlets of trees with their leaves.
Spring wood –n. Wood produced early in the year, characterized by larger ducts and cells than the later growth.
Spumose – a. Frothy.
Spur – n. A short, compact branch with little or no internodal development; a tublar or saclike projection from a blossom, as of a petal or sepal.
Squama (pl. **squamae**) – n. A scale of any sort, usually the homolog of a leaf.
Squamaceous – a. Scaly.
Squamate–a. Furnished with scales, scalelike leaves, or bracts.
Squamellate – a. With small or secondary scales.
Squamule – n. The hypogynous scale of grasses, the lodicule.
Squamulose – a. With small scales.
Squarrose – a. Spreading or recurved at the tip, said of the tips of some lemmas.
Squarrulose – a. Diminutively squarrose.
Stalk – n. The stem of any organ, as the petiole, peduncle, pedicel, filament, culm, or stipe.
Stamen – n. The pollen-bearing organ of the flower, the male organ in the angiosperms.
Stamen, sterile – n. A body belonging to the series of stamens but without pollen.
Staminate – a. Having stamens and no pistil, male.
Stamineal – a. Relating to or consisting of stamens.
Staminode (**staminodium**, pl. **staminodia**) – n. A sterile stamen, or a structure resembling such and borne in the staminal part of the flower; in some flowers, as in *Canna*, staminodia are petal-like and showy.
Standard – n. The upper and broad more or less erect petal of a papilionaceous flower.
Stasad – n. A plant of stagnant water.
Stasimorphy–n. Alternation of form caused by arrested development.
Stasophilous – a. Dwelling in stagnant water.
Staurigamia – n. Cross-fertilization.
Stegium – n. Threadlike appendages sometimes found covering the style of asclepiads.
Stelipilous – a. With stellate hairs.
Stellate – a. Star-shaped or radiating like the points of a star.
Stellate scales – pl. n. Discs borne by their edge or center.
Stelliform – a. Star-shaped.
Stem – n. The main axis of a plant; leaf-bearing and flower-bearing as distinguished from the root-bearing axis.
Stem, subterranean – n. A rhizome, tuber, bulb, or corm.
Stenopetalous – a. Narrow-petaled.
Stenophyllous – a. Narrow-leaved.
Stenotropism – n. The state of having narrow limits of adaptations to varied conditions.
Stereotropism – n. The act of responding to contact stimuli.
Sterile – a. Barren, as a flower destitute of pistil; used for staminate or neuter flowers.
Sternotribal – a. Said of flowers whose anthers are so arranged as to dust their pollen on the under part of the thorax of their insect visitors.

Stigma (pl. **stigmas** or **stigmata**) – n. The part of a pistil or style which receives the pollen; a point on the spores of Equisetum.
Stigmatic – a. Pertaining to the stigma.
Stinging hair – n. A hollow hair seated on a gland which secretes an acid substance, as in nettles.
Stipe– n. The "leaf-stalk" of a fern; the support of a gynoecium or carpel.
Stipitate – a. With a stipe.
Stipular – a. Having stipules or relating to them.
Stipulate – a. Furnished with stipules.
Stipule – n . One of the pair of appendages borne at the base of the leaf in many plants.
Stipulose – a. Having stipules.
Stolon – n. A sucker, runner, or any basal branch which is disposed to root.
Stoloniferous – a. With stolons or runners that take root.
Stoma (pl. **stomata**) – n. A specialized orifice in the epidermis communicating with intercellular spaces.
Stomate – n. See Stoma.
Stone – n. The hard endocarp of a drupe, a pit.
Stone fruit – n. A drupe such as a plum or peach.
Stool – n. The base of a plant from which offsets or layers are taken; several stems arising from the same root, as in wheat.
Stopple – n. A projection or lid on a pollen grain which falls away to admit passage of the pollen tube.
Stramineous–a. Strawlike or straw-colored.
Strap – n. The ligule of a ray floret in *Compositae*.
Strap-shaped – a. Ligulate or lorate.
Straw – n. The jointed hollow culm (stem) of grasses.
Streptocarpous – a. With fruits spirally marked; with twisted fruits.
Striate – a. With fine grooves, ridges or lines of color.
Strict – a. Stiffly upright, rigid, erect.
Striga – n. A small, straight, hair-like scale.
Strigose – a. Beset with sharp-pointed appressed straight and stiff hairs or bristles; hispid.
Strobilaceous – a. Relating to or resembling a cone.
Strobile – n. An inflorescence made up largely of imbricated scales, as the hop or the pine; a cone.
Strombuliform – a. Spirally twisted like some shells.
Strombus – n. A spirally coiled legume, as in *Medicago*.
Strombus-shaped – a. Shaped like a snail shell.
Strophiole – n. An appendage at the hilum of certain seeds; a caruncle.
Stylar – a. Relating to the style.
Stylar column – n. The column of orchids.
Style – n. The more or less elongated part of the pistil between the ovary and the stigma.
Stylopod (**-ium**) – n. The enlargement at the base of the styles in *Umbelliferae*.
Suaveolent–a. Sweet-smelling, fragrant.
Subcordate – a. Almost cordate.
Suber – n. Cork.
Sub-erose – a. Slightly gnawed in appearance.
Suberous – a. Corky in texture.
Subfamily – n. A group of genera within a family.
Submerged – a. Growing under water.
Suborder – n. A group of genera lower than an order and higher than a family.
Subpetiolar – a. Under the petioles, as the buds in *Platanus*.

Subspecies – n. A group of forms ambiguous in rank, and between a variety and a species.

Subtend – v. To stand below and close to, as a bract below a flower or a leaf below a bud.

Subterraneous – a. Under the ground.

Subtribe – n. A divsion between a tribe and a genus.

Subula – n. A fine, abrupt, sharp point.

Subulate – a. Awl-shaped.

Succulent – a. Juicy or pulpy.

Sucker – n. A shoot of subterranean origin; a haustorium, sometimes restricted to the penetrating organ or papilla.

Suffrutescent – a. Obscurely or somewhat shrubby.

Suffruticose – a. Pertaining to a low and somewhat woody plant; diminutively shrubby or fruticose; woody at base.

Sulcate – a. Grooved or furrowed lengthwise.

Sulcus – n. A small groove.

Sulfureous – a. Sulfur-colored.

Sulphur rain – n. Pine pollen carried in excessive amount by air currents.

Super-axillary – a. Borne above the axil.

Superficiales – pl. n. Said of leptosporangiate ferns, with sori arising from the surface of the frond.

Superior – a. Growing or placed above; also in a lateral flower on the side next to the axis; the posterior, or upper lip of a corolla is the superior.

Superior ovary – n. An ovary with the floral envelopes inserted below it on the torus; a hypogynous ovary.

Supernatant – a. Floating on the surface.

Superposed – a. Placed vertically over some other part.

Supine – a. Prostrate with face turned upward.

Supra-axillary – a. See Super-axillary.

Surculose – a. Producing suckers.

Surculum – n. The rhizome of a fern.

Surcurrent – a. Having winged expansions from the base of the leaf prolonged up the stem.

Suspended ovule – n. An ovule hanging from the apex of the cell.

Suture – n. A junction or seam of union; a line of opening or dehiscence.

Sword-shaped – a. Ensiform.

Syconium – n. A hollow multiple fruit, as that of a fig.

Sylva – n. An account of the trees of a district, or a discourse on trees.

Sylvestrine – a. Of or pertaining to woods.

Symbiont – n. Organism in symbiosis.

Symbiosis – n. The living together of dissimilar organisms, with benefit to one only, or both; also styled commensalism, consortism, individualism, mutualism, nutricism, prototrophy, and syntropism.

Symbiosis, antagonistic – n. A struggle between two organisms.

Symbiosis, conjunctive – n. An intimate blending of the symbionts so as to form an apparently single body.

Symbiosis, disjunctive – n. Symbiosis with no direct union between the symbionts.

Symbiotic – a. Relating to symbiosis.

Symmetrical – a. Actinomorphic; regular, capable of division by a longitudinal plane into similar halves.

Symmetry – n. A due proportion of the several parts of a body to each other.

Symmetry, bilateral - n. Symmetrical arrangement capable of equal division in one plane only, as in a pea flower.

Symmetry, radial - n. Symmetrical arrangement capable of equal division in more than one direction through the center; as in the mallow flower.

Sympatric - a. Inhabiting one and the same area.

Sympetalous - a. With partially or wholly fused petals.

Symphyllode - n. A cone scale of *Abietineae.*

Symphyllodium - n. The combined ovuliferous scales in the flower of certain *Coniferae.*

Symphysis - n. Coalescence; fusion of like parts.

Synacmy - n. A condition in which stamens and pistils mature together; the opposite of heteracmy.

Synantherous - a. With anthers joined together.

Synanthesis - n. The simultaneous maturing of stamens and pistils.

Syncarp - n. A multiple or fleshy aggregate fruit, as the mulberry or magnolia.

Syncarpous - a. Composed of two or more united carpels.

Synchronic species - pl. n. Species which belong to the same time level, contemporary.

Syncolliphytum - n. A plant in which the perianth becomes combined with the pericarp.

Synema (pl. **synemata**) - n. The column of monadelphous stamens, as in *Malvaceae.*

Synoecious - a. Having staminate and pistillate flowers both present in the same head.

Synonym - n. A name with the same meaning as another name in the same language but spelled differently; in taxonomy, synonyms are two or more scientific names for the same taxon, one of which is correct and the others incorrect under the International Rules of Nomenclature.

Synonymous - a. Having the same meaning.

Synonymy - n. Discarded names for identical objects.

Syntype - n. One of two or more specimens or elements used by the author when no holotype was designated, or one of two or more specimens simultaneously designated as type.

Synzoochory - n. Dispersion by animals.

Syrtidophilous - a. Dwelling on dry sand bars.

Systellophytum - n. A plant with a persistent calyx appearing to form part of the fruit.

System - n. A scheme of classification, as the Natural System.

Systematic botany - n. A study of plants in their mutual relationships and taxonomic arrangement.

Systerophyte - n. A plant which lives on dead matter; a saprophyte.

T

Tabasheer - n. A siliceous concretion occurring in the joints of bamboo.

Tachyspore - n. A plant which quickly disperses its seeds.

Tactile - a. Sensitive to the touch.

Taphrophilous - a. Ditch-dwelling.

Tap-root - n. The primary descending root, forming a direct continuation from the radicle.

Tartarous - a. With a loose or rough crumbling surface, as some lichens.

Tassel - n. The staminate inflorescence in maize.

Tawny - a. Fulvous, dull brownish-yellow.

Taxis – n. The reaction of free organisms in response to external stimuli by movement.

Taxon (pl. **taxa**) – n. A general term applied to any taxonomic element, population, or group, irrespective of its classification level.

Taxonomy – n. The systematic classification of organisms.

Tectoparatype – n. A specimen selected to show the microscopic structure of the original type of a species or genus.

Tectum – n. Roof.

Tegule – n. One of the involucral bracts subtending the flower head in *Compositae*.

Teknospore – n. A spore produced directly from male or female organs of *Equisetaceae* and many ferns.

Teleology – n. The doctrine of final causes, or theory of tendency to an end.

Telmatophilous – a. Marsh-loving.

Telmicolous – a. Dwelling in fresh-water marshes.

Telotropism – n. The act of turning to one stimulus to the exclusion of all others.

Temulentous – a. Drunken, nodding in a jerky, irregular manner.

Tendril – n. A rotating or twisting threadlike process or extension by which the plant grasps an object and clings to it for support; morphologically it may be a modified stem, leaf, leaflet, or stipule.

Tentacle – n. A sensitive glandular hair, as those on the leaf of *Drosera*.

Tepal – n. Used in the plural for sepals and petals of similar form and not readily differentiated.

Teratology – n. The study of malformations and monstrosities.

Terete – a. Circular in transverse section.

Terminal – a. Proceeding from, or belonging to the end or apex.

Ternary – a. In threes, trimerous; the result of the third axial order, as derived from the primary.

Ternate – a. In threes.

Terrestrial – a. Growing in the soil in distinction from growing in water or other habitats.

Terricolous – a. Dwelling in the ground.

Tesselate – a. With the surface marked with square or oblong depressions.

Testa – n. The outer coat of the seed, usually hard and brittle.

Testiculate – a. Shaped like the tubers of orchids and fruit of *Mercurialis*.

Tetrad – n. A group of four objects, as the four pollen-grains formed from one pollen-mother cell.

Tetradymous – a. Having four cells or cases.

Tetradynamous – a. Having four long stamens and two short, as in *Cruciferae*.

Tetragonal – a. Four-angled.

Tetramerous – a. Of four members.

Tetrandrous – a. With four stamens.

Tetrapetalous – a. With four petals.

Tetrapterous – a. Four-winged.

Tetrasepalous – a. With four sepals.

Tetrastachyous – a. With four spikes.

Tetrastichous – a. In four vertical ranks.

Thalamous – n. The receptacle of a flower.

Thalloid – a. Resembling or shaped like a thallus.

Thallus – n. A flat leaflike organ; in some cryptogams, the entire cellular plant body without differentiation into stem and foliage.

Thelephorous – a. Covered with nipple-like prominences.

Theoretic diagram – n. A floral diagram of the theoretic components, not necessarily the same as seen on inspection.

Thermocleistogamy – n. Self-pollination taking place with flowers the opening of whose perianth has been inhibited by low temperature.

Thermophilous – a. Dwelling in warm waters.

Thermotropism – n. The act of turning in response to heat.

Therophyllous–a. Producing leaves in summer, deciduous-leaved.

Therophyte – n. A plant which completes its development in one season, its seeds remaining latent during the hot season; an annual.

Thigmomorphosis – n. Change in the original structure due to contact, as the adhering discs of *Ampelopsis*.

Thigmotaxis – n. The result of mechanical stimulus.

Thigmotropism – n. The act of turning in response to a mechanical stimulus.

Thinicolous – a. Dwelling on shifting sand dunes.

Thinophilous – a. Dune-loving.

Thorn – n. A spine, usually an aborted branch, simple or branched.

Throat – n. The opening or orifice into a gamopetalous corolla or perianth; the place where the limb joins the tube.

Thyrse – n. A compact and more or less compound panicle; more correctly a panicle-like cluster with main axis indeterminate and other parts determinate.

Thyrsoid – a. Resembling a thyrse.

Thyrsula – n. A little cyme which is borne by most labiates in the axil of the leaves.

Tiller – n. A sucker or branch from the base of the stem.

Tillering – a. Throwing out shoots from the base of the stem.

Timber-line – n. The upper limit of tree vegetation on the mountains or high latitudes.

Tiphad – n. A pond plant.

Tiphophilous – a. Pond-loving.

Tomentose – a. With tomentum; densely woolly or pubescent; with matted soft woollike hairiness.

Tomentulose – a. Somewhat or delicately tomentose.

Tomentum – n. A densely matted pubescence.

Tongue – n. Ligule.

Tooth – n. A small, pointed marginal lobe.

Toothed – a. Dentate.

Topotropism – n. A turning towards a place from which a stimulus comes.

Topotype – n. A specimen of a named species from the original locality.

Torose – a. Cylindrical with contractions at intervals, somewhat moniliform.

Torulose – a. The diminutive of torose.

Torus – n. The receptacle of a flower, that portion of an axis on which the parts of the flower are inserted.

Trabecular – a. Like a cross-bar.

Trabeculate – a. Cross-barred.

Trace – n. A strand of vascular tissue connecting the leaf with the stem.

Trachycarpous – a. Rough-fruited.

Trailing – a. Prostrate, but not rooting.

Trap flowers – pl. n. Prison flowers, which confine insect visitors until pollination has taken place.

Trap–hairs – pl. n. Special hairs which confine insects in certain flowers until pollination is effected.

Trap-prison – n. A flower, such as *Aristolochia,* which confines insect visitors until pollination has taken place.
Traumatic – a. Of or pertaining to a wound.
Traumatism – n. Abnormal growth in consequence of injury.
Traumatropism – n. The sensitiveness of certain plant organs to wounds.
Tree – n. A woody plant that produces one main trunk and a more or less distinct and elevated head
Triachaenium – n. A fruit similar to a cremocarp, but of three carpels.
Triad – n. A group of three objects.
Triadelphous – a. With stamens in three sets.
Triandrous – a. Having three stamens.
Triangulate – a. Three-angled.
Tribe – n. A group superior to a genus, but less than an order.
Tricamarous – a. Said of a fruit composed of three loculi.
Tricarinate – a. With three keels or angles.
Tricarpellary – a. Of three carpels.
Trichasium – n. A cymose inflorescence with three branches.
Trichocarpous – a. Hairy-fruited.
Trichocephalous – a. With flowers collected into heads, and surrounded by hairlike appendages.
Tricholoma (pl. **tricholomata**) – n. An edge or border with hairs.
Trichome – n. Any hairlike outgrowth of the epidermis, as a hair or bristle.
Trichotomous – a. Three-forked, branching into three divisions.
Tricolor – a. Three-colored.
Tricussate – a. Said of whorls of three leaves each, ternate.
Tridentate – a. Three-toothed.
Tridigitate – a. Thrice digitate, with three fingers.
Tridynamous – a. With three stamens out of six being longer than the rest.
Trifid – a. Divided into three parts.
Trifoliate – a. Having three leaflets.
Trifurcate – a. Having three forks or branches.
Trigamous – a. Bearing three kinds of flowers, trimorphic.
Trigeminous – a. Tergeminate, trijugate, triple .
Trigonous – a. Three-angled.
Trilobate – a. Three-lobed.
Trimerous – a. In threes, three-membered parts.
Trimonoecious – a. With perfect, staminate and pistillate flowers on the one plant.
Trimorphic – a. Occurring under three forms, as with long, short, and intermediate styles.
Trimorphism – n. Heterogamy, with long, short, and mid-styled flowers.
Trinervate – a. Three-nerved.
Trioecious – a. With perfect, staminate and pistillate flowers on different individual plants within the species.
Tripartite – a. Divided into three parts.
Tripinnate – a. Thrice pinnate.
Tripterous – a. Three-winged.
Triquetrous – a. With three salient angles.
Tristichous – a. In three vertical rows.
Trisulate – a. With three grooves or furrows.
Triternate – a. Three times three; the leaflets or segments of a twice ternate leaf again divided into three parts.
Tropic – a. Reacting to a stimulus by external change in an organism.
Tropism – n. A curvature which results from a response to some stimulus; the disposition to respond by turning or bending.

Trumpet-shaped - a. Tubular with dilated orifices, salver-shaped.

Truncate - a. Ending abruptly, the base or apex nearly or quite straight across.

Tryma (pl. **trymata**) - n. A drupaceous nut with dehiscent exocarp.

Tuber-n. A short, thickened branch of a subterranean stem, beset with buds or "eyes."

Tubercle - n. A little tuber; any excrescence, as on the roots, ascribed to the action of symbiotic organisms.

Tuberculate - a. Furnished with knoblike excrescences or tubercles.

Tuberiferous - a. Bearing tubers.

Tuberoid - a. Said of a fleshy-thickened root, resembling a tuber, as in many terrestrial orchids.

Tuberous - a. With tubers, tuber-like.

Tufted - a. Cespitose, clustered, or clumped.

Tumescent - a. Somewhat tumid, inflated, or swollen.

Tumid - a. Swollen.

Tunic - n. The skin of a seed, the spermoderm; the coat of a bulb; any loose membranous skin not formed from the epidermis.

Tunicated - a. Composed of concentric layers or coats, as the bulb of an onion.

Turbinate - a. Top-shaped.

Turfaceous - a. Pertaining to bogs.

Turgescence - n. The distension of a cell or cellular tissue by water or other liquid.

Turgid - a. Swollen from fullness, but not from air.

Turion - n. A scaly, often thick and fleshy, shoot produced from a bud on an underground root-stock, as *Asparagus*.

Tussock - n. A tuft of grass or grasslike plants.

Twig - n. A small shoot or branch of a tree.

Twiner - n. A plant which twines or climbs by winding its stem around a support.

Type - n. A nomenclatural type is that constituent element of a taxon to which the name of the taxon is permanently attached, whether as an accepted name or as a synonym.

Type specimen - n. The original specimen from which a description was written.

Type specimens - pl. n. Icotypes.

Typical - a. In classification, conforming to the originally described specimen.

Typonym - n. A synonym; a name based on the same type, specimen, or concept as another and older name.

U

Ubiquitous - a. Occurring everywhere.

Uliginose - a. Growing in swamps.

Umbel - n. An indeterminate inflorescence consisting of several pedicellate flowers having a common point of attachment.

Umbel, compound - n. An umbel with each ray itself bearing an umbel.

Umbel, cymose - n. An apparent umbel, but with the flowers opening centrifugally; a cyme which simulates an umbel.

Umbellate - a. Umbelled; with umbels; pertaining to umbels.

Umbellet - n. A secondary umbel.

Umbelliferous - a. Bearing umbels.

Umbelliform - a. In the shape of an umbel.

Umbellule - n. An umbellet; a small umbel.

Umbilical cord - n. A vascular strand by which seeds are sometimes attached to the placenta, the funiculus.

Umbilicate – a. Depressed in the center.

Umbilicus – n. The hilum of a seed.

Umbo (pl. **umbones**) – n. A boss or protuberance.

Umbonate – a. Bearing an umbo or boss in the center.

Umbonulate – a. Having or ending in a very small boss or nipple.

Umbracticolous – a. Growing in shady places.

Umbraculiferous – a. Bearing an umbrella.

Unarmed – a. Destitute of prickles or other armature; sometimes it means pointless, muticous.

Uncate – a. Hooked, bent, the tip in the form of a hook.

Uncinate – a. Hooked at the point, with hooks.

Unctuous – a. Having a surface which feels greasy.

Undate – a. Wavy, undulate.

Undulate – a. Wavy, repand.

Unguicular – a. Furnished with a claw.

Unguiculate – a. Contracted at the base into a claw.

Ungulate – a. Clawed.

Unicarpellate – a. With fruit consisting of a single carpel.

Unilateral – a. One-sided, either originating on or, usually, all turned to one side.

Unipetalous – a. Having a corolla of only one petal, the others not being developed (not used for gamopetalous).

Uniseriate – a. In one horizontal row or series.

Unisexual – a. Of one sex; with either stamens or pistil or their representative.

Unitypic – a. Monotypic, of one type.

Unsymmetrical – a. Irregular, asymmetrical.

Urceolate – a. Pitcher-like, hollow and contracted at the mouth like an urn or pitcher.

Urceolus – n. The two confluent bracts of *Carex,* the utricle; any flask-shaped anomalous organ; small pitcher.

Urn – n. The base of a pyxis.

Utricle – n. A small bladdery pericarp, as in *Atriplex;* a membranous sac surrounding the fruit proper in *Carex*; any bladder-shaped appendage.

Utricular – a. Bladder-shaped.

V

Vagina – n. The sheathing petiole which forms a continuous tube, as in sedges or grasses.

Vaginate – a. Sheathed.

Vaginiferous – a. Bearing sheaths.

Vallecula – n. The grooves in the intervals between the ridges in the fruit of *Umbelliferae.*

Vallecular – a. Of or pertaining to the grooves in the fruit of *Umbelliferae.*

Valvate – a. Opening by valves, as in most dehiscent fruits and some anthers; parts of a flower bud that meet without overlapping.

Valve – n. One of the pieces into which a capsule naturally separates at maturity; the segment of a calyx meeting in vernation without overlapping; a partially detached flap of an anther.

Variant – n. A form arising from variation.

Varicose – a. Abnormally enlarged in places, irregularly swollen.

Variety – n. A group of organisms within a species that differs from other members or groups within the species in one or more minor characteristics but not enough to justify a new specific epithet.

Vascular – a. With vessels or ducts.

Vasculum – n. A collecting can for botanical specimens.

Vegetable – a. Belonging to or consisting of plants.

Vegetation - n. The sum total of all plants growing on an area.

Vein - n. A strand of vascular tissue in a flat organ such as a leaf.

Velamen (pl. **velamines**) - n. A parchment-like sheath or layer of spiral-coated air cells on the roots of some tropical epiphytic orchids and aroids.

Velum - n. The membranous indusium in *Isoëtes*.

Velumen - n. Close, short, soft hairs.

Velutinous - a. Velvety, due to a coating of fine soft hairs.

Venation - n. Veining; arrangements or disposition of veins.

Venenose - a. Very poisonous, venomous.

Venomous - a. Poisonous.

Ventral - a. Of or pertaining to the belly; pertaining to or designating that surface of a carpel, petal, etc., which faces toward the center of the flower.

Ventural suture - n. The ventral seam or line of dehiscence in a carpel.

Ventricose - a. Swelling or inflated on one side, as the corolla of some labiates and *Scrophulariaceae*.

Venuloso-hinoideous - a. Said of veins which proceed from the midrib and are parallel and cross-veined.

Vermicular - a. Worm-shaped.

Vermiform - a. Worm-shaped.

Vernal - a. Pertaining to spring.

Vernalization - n. The process of shortening the vegetative period of plants by seed treatment.

Vernation - n. The disposition or arrangement of leaves in the bud.

Vernicose - a. Shiny as though varnished.

Verrucose - a. Covered with wart-like elevations.

Versatile - a. Hung or attached near the middle and usually moving freely, as an anther attached crosswise on the apex of a filament and capable of turning.

Versicolor - a. Variously colored, as of one color blending into another, or changing in color.

Verticil - n. A whorl, or circular arrangement of similar parts round an axis.

Verticillaster - n. A false whorl, composed of a pair of opposed cymes, as in *Labiatae*.

Verticillastrate - a. Bearing or arranged in clusters resembling whorls.

Verticillate - a. Whorled with two or more leaves at a node, cyclical.

Vesicle - n. A small bladder or cavity.

Vesicular - a. Composed or covered with little bladders or blisters.

Vespertine - a. Appearing or expanding in the evening.

Vestige - a. The remaining trace of an organ which was fully developed in some ancestral form.

Vestigial - a. Rudimentary.

Vexillum - n. The standard or large posterior petal of a papilionaceous flower.

Viatical - a. Growing by roadsides or paths.

Villose - a. With long, silky, straight hairs.

Vimineous - a. Bearing long and flexible twigs.

Vinaceous - a. Wine-colored, purplish-red.

Vine - n. The plant which bears grapes, *Vitis vinifera*; in the U.S. applied to any trailing or climbing stem, or runner.

Vinicolor - a. The color of wine, dark or purple-red.

Viniferous - a. Wine-bearing.

Violaceous - a. Violet-colored, ianthine.

Virescent – a. Turning green.

Virgate – a. Wand-shaped, twiggy.

Viridescent – a. Becoming green.

Viscid – a. Sticky from a tenacious coating or secretion.

Vitreous – a. Transparent, hyaline, formerly used for the light green of glass.

Vitta – n. An aromatic oil tube of the pericarp of most *Umbelliferae.*

Viviparous – a. Germinating or sprouting from seed or bud while attached to the parent plant.

Volute – a. Rolled up in any way.

W

Wart – n. A hard or firm excrescence.

Web – n. A network of interlacing threads or fibers.

Wedge-shaped – a. Cuneate.

Weed – n. A plant detrimental to man's interest, displeasing to the eye, or of no apparent value.

Weedy – a. With the attributes ot a weed.

Weel – n. An arrangement of hairs which keeps out unbidden insect guests from flowers.

Whorl – n. Cyclic arrangement of appendages at a node.

Wind-pollinated – a. With the pollen conveyed by the agency of air; anemophilous.

Wing – n. Any membranous expansion attached to an organ; the lateral petal of a papilonaceous flower.

Witches' broom – n. A disease shown by tufts of shoots, due to attack by fungi or mites.

Wood – n. The lignified portion of plants, included within the cambium, but exclusive of the pith.

Woolly – a. Lanate, tomentose, clothed with long and tortuous or matted hairs.

Wrinkled – a. Rugose, creased.

X

Xanthic – a. Tending toward yellow.

Xanthophyll – n. A yellow substance insoluble in water and associated with chlorophyll.

Xanthorrhiza – n. Yellow-root.

Xenia – n. The direct influence of foreign pollen on the parts of the mother plant.

Xenochroma – n. The effect of foreign pollen producing a change in the color of the fruit.

Xenogamy – n. Cross-fertilization.

Xeriobole – n. A plant whose seeds are scattered by dehiscence due to dryness.

Xerochase – n. A fruit that opens in dry air and closes in humid air.

Xerochastic – a. Said of plants whose fruits burst by desiccation, thereby scattering their seeds or spores.

Xerophilous – a. Growing in arid places.

Xerophyte – n. A plant which can subsist with a small amount of moisture, a desert plant.

Xerotropism – n. The tendency of plants, or parts thereof, to alter their position to protect themselves from desiccation.

Xylem – n. The wood elements of a vascular bundle.

Xylem rays – pl. n. – A radial plate of xylem between two medullary rays.

Z

Zenotropism – n. Negative geotropism.

Zigzag – a. Having short bends or angles from side to side.

Zonate – a. Marked circularly as the leaves of *Pelargonium zonale*; zoned, banded.

Zoned – a. Colored in rings or circles.

Zoöchore – n. A plant distributed by animals.

Zoöphilous – a. Pollinated by the agency of animals.

Zoöspore – n. A free-moving spore of the lower cryptogams; an asexual reproductive cell with cilia.

Zygomorphic – a. Capable of division by only one plane of symmetry.

Subject Classification

Aestivation

Aestivation – n. The manner in which the floral parts are arranged in the bud before expansion.

Convolute – a. With petals rolled up in such a way that the outer part of each covers the inner part of the one in front of it, while in turn its inner part is covered by the one behind it. In cross section, the petals resemble curved spokes in a wheel.

Corrugate, or crumpled – a. Characterized by the irregular crumpling of otherwise plane petals due to rapid growth in a confined space.

Imbricate – a. With the outer parts overlapping the inner parts, as the shingles on a roof; they break joints.

Induplicate – a. Valvate with the margins of each part projecting inward.

Involute – a. Valvate with the margins of each part rolled inward.

Plicate – a. With the parts folded lengthwise.

Quincuncial – a. In aestivation, partially imbricated of five parts, two being exterior, two interior, and a fifth one having one margin exterior and the other interior, as in the calyx of the rose.

Reduplicate – a. Valvate with the margins projecting outward.

Valvate – a. With the parts meeting by their abrupt edges without overlapping or turning.

Agents of Pollination

Animal-pollinated:

Zoöchore – n. A plant distributed by animals (usually applied to plant dispersal).

Zoöphilous – a. Animal-loving; flowers pollinated by animals.

Bat-pollinated:

Chiropterophilous – a. Said of flowers pollinated by bats.

Bird-pollinated:

Ornithogamous – a. Said of flowers pollinated by birds.

Ornithophilous – a. Said of flowers pollinated by birds.

Insect-pollinated:

Cantharophilcus – a. Said of flowers pollinated by beetles.

Dipterid – n. Fly-flower. Flowers visited by dipterous flies.

Entomogamous – a. Pollinated by insects.

Entomogamy – n. The pollination of flowers by insects.

Entomophilous – a. Said of flowers dependent upon insects for pollination.

Hover-fly flowers – n. Flowers adapted for pollination by *Syrphidae*.

Humble-bee (bumble-bee) flowers – n. Flowers especially adapted to pollination by the species of *Bombus*.

Insect-pollination – n. The transfer of pollen from the anther to the stigma by insects.

Lepidopterid – n. Flower adapted for pollination by moths and butterflies.

Moth-flowers – n. Flowers adapted for pollination by moths; they are usually white night-bloomers.

Muscarian flowers – n. Flowers that attract flies by a putrid stench.

Necrocoleopterophilous – a. Said of flowers pollinated by carrion beetles.

Psychophilous – a. Said of flowers pollinated by diurnal lepidoptera.

Sapromyiophilous – a. Said of flowers pollinated by carrion- or dung-flies.

Sphingophilous–a. Said of flowers pollinated by hawk-moths and other nocturnal lepidoptera.

Man-pollinated:

Anthropochorous – a. Distributed by man (usually applied to plant dispersal).

Self-pollinated:

Autogamous – a. Characterized by self-fertilization.

Autogamy – n. The fertilization of a flower by its own pollen.

Cleistogamous – a. With small closed self-pollinated flowers.

Close-pollinated – a. Pollinated by its own pollen.

Self-pollination – n. Pollination by its own pollen.

Snail-pollinated:

Malacophilous – a. Said of flowers pollinated by snails and slugs.

Water-pollinated:

Hydrocarpic – a. Said of aquatic plants which are pollinated above the water but withdraw the fertilized flowers below the surface for development, as in *Vallisneria*.

Hydrochore – n. A plant distributed by water (usually applied to plant dispersal).

Hydrophilous – a. Water-loving; said of flowers pollinated by water.

Wind-pollinated:

Anemochorous – a. Wind-distributed (usually applied to plant dispersal).

Anemogamous – a. Wind-pollinated.

Anemophilous – a. Wind-pollinated.

Corolla

Achilary – a. Without a lip, as in some orchids.

Achlamydeous – a. Without a perianth, as in willows.

Actinomorphic–a. With radial symmetry, regular.

Acyclic – a. With the parts arranged spirally, not in whorls.

Aestivation – n. The manner in which the parts of a flower are folded up in the bud.

Ambigenous – a. Said of a perianth whose exterior is calycine and the interior corolline, as in *Nymphaea*.

Amphichromy – n. A display of two different colors when in flower.

Apetalous – a. Without petals.

Banner – n. The standard of a papilionaceous flower.

Bilabiate – a. Two-lipped.

Calcarate – a. Spurred.

Campanulate – a. Bell-shaped.

Cardiopetalous – a. With heart-shaped petals.

Carina – n. A keel; used either for the two combined lower petals of a papilionaceous corolla or for a salient longitudinal projection on the center of the lower face of an organ, as on the lemmas of many grasses.

Cement-disk – n. The retinaculum of orchids.

Chasmogamous – a. Said of a flower whose opening precedes pollination; see *Cleistogamous.*

Chasmogamy – n. The opening of the perianth at flowering time, the opposite of cleistogamy.

Chloranthy – n. The reversion of petals to green leaves.

Choripetalous–a. Polypetalous, with separate petals.

Claw – n. The long narrow petiole-like base of the petals or sepals in some flowers.

Clip – n. The seizing mechanism in the flowers of asclepiads.

Corolla – n. The inner floral envelope composed of separate or connate petals.

Corolline – a. Seated on a corolla; corolla-like; petaloid; belonging to a corolla.

Corona – n. Crown, coronet; any appendage or intrusion that stands between the corolla and stamens, or on the corolla, as the cup of a daffodil, or that is an outgrowth of the staminal part or circle, as in the milkweed.

Cyclic – a. In whorls, not spirals.

Deflorate – a. Past the flowering state.

Dialypetalous – a. With separate petals, polypetalous.

Dichlamydeous – a. With a double perianth, calyx and corolla.

Disk flowers – n. The tubular flowers in the center of heads of *Compositae,* as distinguished from the ray flowers.

Epipetalous – a. Borne upon the petals; placed before the petals.

Euephemerous – a. Said of flowers which open and close within 24 hours.

Faucal – a. Pertaining to the throat of a gamopetalous corolla.

Faux (pl. **fauces**) – n. Usually used in the plural to designate the throat in a gamopetalous corolla.

Floral – a. Pertaining to flowers.

Floral diagram – n. A drawing to show the relative position and number of the constituent parts of a flower.

Floral envelope – n. The perianth leaves, the calyx and corolla.

Florepleno – a. With full or double flowers.

Floret – n. A small flower, usually one of a cluster; in grasses, the flower with the two subtending bracts.

Floscule – n. A little flower, a floret.

Flower – n. A modified plant structure concerned with the production of seeds in angiosperms.

Funnelform – a. With tube gradually widening upward and passing insensibly into a limb, as in many flowers of *Convolvulus;* infundibuliform.

Galea – n. A petal shaped like a helmet, placed next to the axis, as in *Aconitum.*

Galeate – a. Hollow and vaulted as in many labiate corollas.

Gamopetalous – a. With petals united, corolla in one piece.

Gorge – n. The throat of a flower.

Haplochlamydeous – a. Monochlamydeous, having a single perianth.

Hemeranthous – a. Day-flowering.

Hercogamy – n. Applied to hermaphrodite flowers in which some structural peculiarity prevents self-pollination.

Hypochil - n. The (often fleshy or otherwise modified) basal portion of the labellum or lip in *Orchidaceae*.

Hypocrateriform - a. Salverform.

Infundibuliform - a. Funnelform.

Keel - n. The united lower petals of a papilionaceous flower.

Labellum - n. The third petal of orchids, usually enlarged and by torsion of the ovary becoming anterior from its normal posterior position; a lip.

Labiate - a. Lipped; of or pertaining to the *Labiatae*.

Labium - n. The lower lip of a labiate flower.

Lepanthium - n. A petal that contains a nectary.

Ligule - n. A strap-shaped body such as the limb of the ray florets in *Compositae*; the lobe of the outer corona in *Stapelia*.

Limb - n. A border, the expanded part of a gamopetalous corolla, as distinct from the tube or throat; the lamina of a petal.

Lip - n. One of the two divisions of a bilabiate corolla or calyx, that is, a gamopetalous corolla cleft into an upper and lower portion; the labellum of orchids.

Mitra - n. The galea of a corolla.

Monochlamydeae - pl. n. A large division of phanerogams which have only one set of floral envelopes.

Monochlamydeous - a. With only one set of floral envelopes.

Monopetalous - a. One-petaled; gamopetalous, with the corolla composed of several petals laterally united.

Monosymmetrical - a. Capable of being dissected equally in one plane only; zygomorphic; bilaterally symmetrical.

Palate - n. In personate corollas, a rounded projection or prominence of the lower lip closing the throat or very nearly so.

Peloria, pelory - n. Reversion, on the part of the individual, to the production of regular flowers, when the species typically has asymmetrical or bilaterally symmetrical flowers.

Perianth - n. The floral envelope, calyx and corolla.

Personate - a. Said of a bilabiate corolla having a prominent palate.

Petal - n. One of the leafy expansions in the floral whorl styled the corolla.

Petaliferous - a. Petal-bearing.

Petalode - n. An organ simulating a petal.

Petaloid - a. Like a petal, or having a floral envelope resembling petals.

Pitfall flowers - n. Transitional flowers, such as *Asarum*, which detain small *Diptera*.

Pleiopetalous - a. Many-petaled.

Pleiopetaly - n. Doubleness in flowers.

Polypetalous - a. With several distinct petals.

Protanthesis - n. The normal first flower of an inflorescence.

Quincuncial - a. Arranged in a quincunx; in aestivation, partially imbricated of five parts, two being exterior, two interior, and the fifth one having one margin exterior and the other interior, as in the calyx of the rose.

Rictus - n. The mouth or gorge of a bilabiate corolla.

Ringent - a. Gaping, as the mouth of an open bilabiate corolla.

Rotate - a. Wheel-shaped, circular and flat, applied to a gamopetalous corolla with a short tube.

Salverform - a. With a slender tube and an abruptly expanding limb, as that of the phlox; hypocrateriform.

Schizopetalous - a. With cut petals.

Seasonal amphichromatism – n. The production of two differently colored flowers on the same stock due to season.

Seasonal heterochromatism – n. Different colors in the flowers of the same inflorescence due to season.

Standard – n. The upper and broad more or less erect petal of a papilionaceous flower.

Stenopetalous – a. Narrow-petaled.

Symmetrical – a. Actinomorphic, regular, capable of division by a longitudinal plane into similar halves.

Throat – n. The opening or orifice into a gamopetalous corolla or perianth; the place where the limb joins the corolla tube.

Trap flowers – n. Prison flowers which confine insect visitors until pollination has taken place.

Trumpet-shaped – a. Tubular with a dilated orifice, salverform, hypocrateriform.

Unilabiate – a. One-lipped.

Vexillum – n. The standard or large posterior petal of a papilionaceous flower.

Dehiscence

Assumentum (pl. **assumenta**) – n. One of the two valves of a silique.

Circumscissile – a. Opening or dehiscing by a horizontal line around the fruit or anther.

Dehisce – v. To open spontaneously when mature, as seed capsules.

Dehiscence – n. The method or process of opening of a seed-pod or anther.

Dehiscent – a. Said of that which dehisces, as the opening of a fruit or anther along lines of suture.

Fissile – a. Tending to split or easily split.

Hydrochastic – a. Said of plants in which the bursting of the fruit and the dispersion of the seeds are caused by absorption of water.

Locucidal – a. With dehiscence on the back between the partitions into the cavity.

Operculate – a. Opening by a lid.

Porocidal – a. Opening by pores.

Ruptile – a. Dehiscing in an irregular manner.

Septicidal – a. With dehiscence along lines of union of the carpels.

Septifragal – a. With the valves breaking away from the dissepiments in dehiscence.

Suture – n. A junction or seam of union; a line of opening or dehiscence.

Ventral suture – n. The ventral seam or line of dehiscence in a carpel.

Xerochase – n. A fruit that opens in dry air and closes in humid air.

Xerochastic – a. Said of plants whose fruits burst by desiccation, thereby scattering their seeds or spores.

Direction

Amphigean – a. Native around the world.

Austral – a. Southern.

Boreal – a. Northern.

Deflexed – a. Bent or turned abruptly downward.

Dextrorse – a. Turning to the right, clockwise.

Eutropic – a. Twining with the sun, clockwise, dextrorse.

Geonasty – n. The act of curving toward the ground.

Geotropic – a. Turning toward the earth.

Hesperal – a. Of the West.

Homalotropous – a. Said of organs which grow in a horizontal direction.

Hyperboreal - a. Of the far North.

Impressed - a. Bent inward, hollowed or furrowed as if by pressure.

Meridional - a. Southern (in the Northern Hemisphere).

Occidental - a. Western.

Oriental - a. Eastern.

Parallel - a. Extended in the same direction, but equally distant at every point.

Porrect - a. Directed outward and forward.

Septentrional - a. Northern.

Sinistrorse - a. Turning to the left or counterclockwise.

Zigzag - a. Having short bends or angles from side to side.

Dispersal

Aelophilous - a. Disseminated by wind.

Anemochore - n. An organism that is disseminated by the wind.

Anemophilous - a. Distributed by wind.

Anthropophilous - a. Plants which follow man.

Blastochore - n. A plant distributed by offshoots or buds.

Bolochore - n. A plant distributed by propulsion.

Bradyspore - n. A plant which disperses its seeds slowly.

Brotochore - n. A plant dispersed by man.

Centrospore - n. A plant with spiny disseminules.

Clitochore - n. A plant which is distributed by falling or sliding.

Disseminule - n. A seed, fruit, or spore modified for dispersal.

Edobole - n. A plant whose seeds are scattered by propulsion through turgescence.

Glacospore - n. A plant with viscid disseminules.

Hydrochore - n. A plant distributed by water.

Hydrophilous-a. Water-loving; distributed by water.

Migration - n. Any movement by which the range of a species is extended (strictly speaking, it means moving under its own power.)

Migrule - n. The unit of migration, as seed, fruit, runner, bulb, etc.

Ornithophilous - a. Bird-loving; distributed by birds.

Sarcospore - n. A plant with fleshy disseminules.

Saurochore - n. A plant disseminated by lizards or snakes.

Synzoöchory - n. Dispersion by animals.

Xeriobole - n. A plant whose seeds are scattered by dehiscence due to dryness.

Zoöphilous - a. Distributed by animals.

Fruits

Carpography

Acarpic - a. Without fruit.

Acarpotropic - a. Not throwing off its fruits.

Achene - n. A small, hard, dry, indehiscent one-seeded fruit in which the ovary wall is free from the seed.

Achenodium - n. A double achene, as the cremocarp of *Umbelliferae.*

Aggregate fruit - n. A cluster of ripened ovaries traceable to separate pistils of the same flower and inserted on a common receptacle.

Akene - n. See Achene.

Amphicarpous - a. Producing two kinds of fruit.

Amphore - n. The lower part of a pyxis, as in henbane.

Anthocarpous - a. Said of fruits with accessories, sometimes pseudocarps, as in the strawberry and pineapple.

Apogamous – a. Developed without fertilization.

Apyrenous – a. Said of fruit which is seedless.

Article – n. The portion of a fruit (especially in *Leguminosae*) separated from others by a constriction or joint, as in *Desmodium.*

Assumentum (pl. **assumenta**) – n. One of the two valves of a silique.

Atrocarpous – a. Black-fruited.

Autocarp – n. A fruit obtained by self-fertilization.

Baccate – a. Berry-like, pulpy or fleshy.

Balausta – n. The fruit of a pomegranate with a firm rind, berried within, and crowned with the lobes of an adnate calyx.

Beak – n. A long prominent and substantial projection; applied particularly to the prolongation of fruits and carpels.

Berry – n. A mature, fleshy, few- to many-seeded. ovary of a single pistil.

Biferous – a. Producing two crops of fruit in one season.

Bilocular – a. Two-celled, with two compartments.

Bivalvular – a. With two valves.

Brachycarpous – a. Short-fruited.

Bradycarpic – a. Fruiting after winter, in the second season after flowering.

Bur – n. Any fruit with a rough or prickly envelope, whether a pericarp, a persistent calyx, or an involucre, as of the sandbur and burdock.

Calyptra – n. A hood or lid; particularly, the hood or cap of the capsule of moss or lid in the fruit of *Eucalyptus.*

Capsella – n. Seed vessel; a small capsule.

Capsular – a. Pertaining to a capsule; formed like a capsule.

Capsule – n. A simple dry fruit, the product of a compound pistil splitting along two or more lines of suture.

Carpel – n. A simple pistil; one unit of a compound pistil; the cone scale in conifers.

Carpography – n. The description of fruits.

Carpophore – n. A portion of receptacle prolonged between the carpels as in *Umbelliferae.*

Caryopsis – n. A small, dry, indehiscent fruit in which the seed coat is adherent to the ovary wall.

Censer-action – n. The action of capsules which like censers (incense-burners), partially open by valves, the seeds being gradually shaken out by wind, as in *Papaver* and *Stramonium.*

Circumscissile – a. Opening or dehiscing along a horizontal line around the fruit or anther, the valve usually coming off like a lid.

Cochlea – n. A closely coiled legume.

Cochleate – a. Spiral, like a snail shell.

Coelospermous – a. Hollow-seeded; said of the seedlike carpels of *Umbelliferae,* with the ventral face incurved at the top and bottom.

Coenocarpium – n. The collective fruit of an entire inflorescence, as a fig or pineapple.

Columella – n. A persistent central axis around which the carpels of some fruits are arranged, as in *Geranium*; the receptacle bearing the sporangia of *Trichomanes* and other ferns.

Commissure – n. The place of joining or meeting, as the face by which one carpel joins another.

Cone – n. A fruit of the pine family *Pinaceae* and of *Cyads*; strobile.

Conelet – n. A little cone, applied to a cone of the first year.

Conocarpium–n. An aggregate fruit consisting of many fruits on a conical receptacle, as the strawberry.

Conoid – a. Conelike.

Creatospore – n. A plant with nut fruits.

Cremocarp – n. A dry, seedlike fruit, composed of two one-seeded carpels invested by an epigynous calyx, separating when mature into mericarps.

Cupule – n. The cup of such fruits as the acorn; an involucre composed of bracts adherent at least by their base.

Cyamium – n. A kind of follicle resembling a legume.

Cynarrhodion – n. A fruit like that of the rose, fleshy, hollow, and enclosing achenes.

Cypsela – n. An achene invested by an adnate calyx, as the fruit of *Compositae*.

Dasycarpous – a. Thick-fruited.

Dialycarpic – a. Having a fruit composed of distinct carpels.

Dicarpellary – a. Composed of two carpels.

Didymous – a. Found in pairs, as the fruits of *Umbelliferae*; divided into two lobes.

Dischisma (pl. **dischismata**) – n. The fruit of *Platystemon*, which divides into longitudinal carpels, each of which again divides transversely.

Dissepiment – n. A partition in an ovary or pericarp caused by the adhesion of the sides or the carpellary leaves.

Dissilient – a. Bursting asunder.

Drupaceous–a. Resembling a drupe, possessing its character, or producing similar fruit.

Drupe – n. A fleshy, one-seeded indehiscent fruit with the seed enclosed in a stony endocarp.

Drupelet – n. One drupe of a fruit made up of aggregate drupes, as in raspberry.

Endocarp – n. The inner layer of the pericarp.

Eriocarpous – a. Woolly-fruited.

Exocarp – n. The outer layer of the pericarp.

Flask – n. The utricle of *Carex*.

Follicle – n. A single carpellate dry fruit dehiscing along one line of suture.

Fructiferous – a. Producing or bearing fruit.

Fructification – n. The act of fruiting.

Fruit – n. A mature ovary or ovaries with or without closely related parts.

Fruit dots – n. The sori of ferns.

Galbulus – n. The fruit of *Taxodium*; a modified cone, the apex of each carpellary scale being enlarged and somewhat fleshy.

Gourd – n. A fleshy, one-celled, many-seeded fruit with parietal placentation.

Gynobase – n. An enlargement or prolongation of the receptacle bearing the ovary.

Hemicarp – n. A half-carpel, a mericarp.

Hesperidium – n. A berry with a tough, leathery rind, as the orange.

Heterocarpous – n. Producing more than one kind of fruit.

Hip – n. The fruit of the rose; technically, a cynarrhodion.

Hygrochastic – a. Said of plants in which the bursting of the fruit and the dispersion of seeds are caused by absorption of water.

Indehiscent – a. Not opening by valves or along regular lines.

Jugum – n. A ridge on the fruits of *Umbelliferae*.

Key fruit – n. The samara of the ash.

Lasiocarpous–a. Pubescent-fruited.

Legume – n. A dry fruit of a simple pistil usually dehiscing along two lines of suture; the fruit of *Leguminosae*.

Locule – n. A compartment or cell of a pistil or anther.

Loment – n. A flattened legume which is constricted between the seeds, falling apart at the constrictions when mature into one-seeded joints.

Lomentaceous – a. Bearing or resembling loments.

Mace – n. The aril of the nutmeg.

Malicorium – n. The rind of a pomegranate.

Mast– n. The fruit of such trees as beech, oak, hickory, etc.

Megasporocarp – n. The developed megasporangium in *Azolla,* finally containing a single perfect megaspore.

Mericarp – n. One of the achene-like carpels or a closed half-fruit of *Umbelliferae.*

Monolocular – a. One-celled, unilocular, applied to ovaries.

Multiple fruit – n. A cluster of ripened ovaries traceable to the pistils of separate flowers, as in mulberry.

Multiseptate – a. With many partitions.

Naucum – n. The fleshy part of a drupe; seed with a very large hilum.

Naucus - pl. n. Certain cruciferous fruits which have no valves.

Nuciferous – a. Bearing or producing nuts.

Nut – n. An indehiscent, usually one-celled, one-seeded fruit (though usually traceable to a compound ovary) with a bony, woody, leathery, or papery wall and in general, partially or wholly enclosed in an involucre or husk.

Nutlet – n. A small nut.

Nux – n. A nut.

Operculum – n. A lid or cover which separates by a transverse line of division, as in a pyxis.

Paracarpous – a. Said of ovaries whose carpels are joined together by their margins only.

Parietal – a. Borne on, or belonging to a wall.

Parthenocarpy – n. The production of fruit without true fertilization.

Parthenogenesis – n. A form of apogamy in which the oösphere develops into a normal product of fertilization without a preceding sexual act.

Pentacamarous – a. With five locules.

Pentacarpellary – a. With five carpels.

Pepo – n. A gourd type of fruit, a one-celled, many-seeded, inferior fruit with parietal placentas and a pulpy interior.

Pericarp – n. The wall of a mature ovary, consisting of an exocarp, a mesocarp, and an endocarp.

Phaenocarpous – a. Having a distinct fruit, with no adhesion to surrounding parts.

Phragma (pl. **phragmata**) – n. A spurious dissepiment in fruits.

Pit – n. The endocarp of a drupe with the enclosed seed.

Placenta – n. The place in an ovary where the ovules are attached.

Placentation – n. The disposition of the placenta.

Plococarpium – n. A fruit composed of follicles arranged around an axis.

Plurilocular – a. With many cells or locules.

Pod – n. A dry dehiscent pericarp; a rather general uncritical term.

Podocarp – n. A stipitate fruit, that is, with the ovary borne on a gynophore.

Polachena – n. A fruit similar to a cremocarp but composed of five carpels.

Pome – n. A fleshy fruit, the product of a compound pistil with the seeds encased within a cartilaginous wall, as in the apple.
Pomiferous – a. Pome-bearing.
Porocidal – a. Opening by pores.
Putamen (pl. **putamines**) – n. The shell of a nut; the hardened endocarp of stone fruits.
Pyrene – n. Nutlet, particularly the nutlet in a drupe.
Pyxis – n. A capsule with circumscissile dehiscence, the upper portion acting as a lid.
Quinquelocular – a. Five-celled.
Regma (pl. **regmata**) – n. A fruit with elastically opening segments or cocci, as in *Euphorbia*; a form of schizocarp.
Repletum – n. A fruit with the valves connected by threads, persistent after dehiscence such as orchids, *Aristolochia*, and some *Papaveraceae*.
Replum – n. A framelike placenta from which the valves fall away in dehiscence, frequently used so as to include the septum of *Cruciferae* in the term.
Rhizocarp – n. A sporangium such as is produced on rootlike processes of members of the *Marsileaceae*.
Ripe – a. Mature, characterized by the completion of an organ or organism for its allotted function.
Samara – n. Winged, achene-like fruit.
Schizocarp – n. A pericarp which splits into one-seeded portions or mericarps.
Schleranthium – n. An achene enclosed in an indurated portion of the calyx tube, as in *Mirabilis*.
Septifragal – a. With the valves breaking away from the dissepiments in dehiscence.
Silicule – n. A short silique, not much longer than broad.
Sillicle – n. The short fruit of certain *Cruciferae*.
Silique – n. The peculiar pod of the *Cruciferae*, two valves falling away from a frame, the replum, on which the seeds grow and across which a false partition is formed.
Simple fruit – n. A fruit which results from the ripening of a single pistil.
Sorose – n. A fleshy multiple fruit, as a mulberry or a pineapple.
Sorus – n. A cluster of sporangia in ferns.
Sphalerocarpum – n. An accessory fruit, as an achene in a baccate calyx-tube.
Sporocarp – n. A receptacle containing sporangia or spores.
Stone – n. The hard endocarp of a drupe.
Stone fruit – n. A drupe such as a plum or peach.
Streptocarpous – a. With fruits spirally marked; with twisted fruits.
Strobile – n. A fruit made up largely of imbricated scales, as in the hop and the pine; a cone.
Strombuliform – a. Said of fruit that is spirally twisted.
Syconium – n. A multiple hollow fruit, as that of a fig.
Syncarp – n. A multiple or fleshy aggregate fruit, as the mulberry and magnolia.
Syncarpous – a. Composed of two or more united carpels.
Syncolliphytum – n. A plant in which the perianth becomes combined with the pericarp.
Systellophytum – n. A fruit in which a persistent calyx appears to form a part of the fruit.
Trachycarpous – a. Rough-fruited.
Triachaenium – n. A fruit similar to a cremocarp but of three carpels.
Tricarpellary – a. Of three carpels.
Trichocarpous – a. Hairy-fruited.

Tryma (pl. **trymata**) – n. A drupaceous nut with dehiscent exocarp.

Unicarpellate – a. With fruit consisting of a single carpel.

Urn – n. The base of a pyxis.

Utricle – n. A small bladdery pericarp, as in *Atriplex*; a membranous sac surrounding the fruit proper in *Carex*; any bladder-shaped appendage.

Vallecula – n. The grooves in the intervals between the ridges in the fruit of *Umbelliferae*.

Valve – n. A segment into which a capsule naturally separates at maturity.

Xerochastic – a. Said of plants in which the bursting of the fruit and dispersion of seeds is caused by desiccation.

Habitats

Terms concerned with the various kinds of habitats.

Alpine

Acrophilous – a. Dwelling in the alpine region.

Alpestrine–a. Nearly alpine, subalpine.

Coryphad – n. An alpine meadow plant.

Subalpine – a. Below alpine, almost alpine.

Bank

Ochthad – n. A bank plant.

Ochthophilous – a. Bank-loving.

Bog

Turfaceus – a. Pertaining to bogs.

Turfophilous – a. Bog-loving, found in bogs.

Clay

Argillaceous – a. Clayey, pertaining to clay, or clay-colored.

Spiladophilous – n. Dwelling in clay.

Cold

Coryphad – n. An alpine meadow plant.

Crymophilous – a. Loving polar regions; inhabiting polar regions.

Frigid – a. Cold, of cold regions.

Dark

Scotophilous – a. Darkness-loving, dwelling in darkness.

Skotophilous – a. See Scotophilous.

Ditch

Taphrophilous – a. Ditch-loving, growing in ditches.

Dry

Cheradophilous – a. Loving dry habitats; dwelling in dry places.

Chersad – n. A plant growing in dry places.

Eremophilous – a. Desert-loving, dwelling in deserts.

Xerophilous – a. Loving dry places, dwelling in dry places.

Dung

Fimetarious – a. Growing on or among dung.

Fimicolous – a. Inhabiting manure.

Earth

Epigeous – a. Above the soil; growing above the soil.

Geophilous–a. Earth-loving; said of plants which fruit underground.

Hypogeous – a. Below the soil; growing or remaining below the soil.

Terricolous – a. Dwelling on the ground.

Field

Agrophilous – a. Loving grain fields.

Campestrine – a. Of or pertaining to fields.

Hemerophilous – a. Loving cultivation, readily cultivated.

Nomad – n. A pasture plant.

Nomophilous – a. Pasture-loving, inhabiting pastures.

Poad – n. A meadow plant.

Forests

Alsad – n. A grove plant.

Ancophilous – a. Loving mountain glens or valleys.

Dendrocolous – a. Dwelling on trees.

Dendrophilous – a. Dwelling on or among trees, loving trees.

Helohylophilous – a. Loving wet forests.

Hylacolous – a. Tree-dwelling.

Hylocolous – a. Inhabiting forests.

Hylodophilous – a. Loving dry woods; dwelling in dry woods.

Nemorose – a. Growing in the woods.

Nemus (pl. **nemores**) – n. Woods.

Orgadophilous – a. Loving open woodland; dwelling in open forests.

Stenophyllophilous – a. Loving deciduous forests.

Sylvatic – a. Growing among the trees.

Sylvestrine – a. Growing in woods.

Gravel

Chalicad – n. A gravel-slide plant.

Chalicodophilous – a. Loving gravel-slides; inhabiting gravel-slides.

Glareose – a. Frequenting gravel.

Hedges

Sepicolous – a. Inhabiting hedges.

Humus

Sapromyiophilous – a. Humus-loving; inhabiting humus.

Lakes or ponds

Lacustrine – a. Belonging to, or inhabiting lakes and ponds.

Lentic – a. Pertaining to, or living in quiet or still water.

Limnophilous – a. Dwelling in lakes.

Tiphad – n. A pond plant.

Tiphophilous – a. Pond-loving; inhabiting ponds.

Limestone

Calcareous – a. Of or pertaining to limestone.

Calcicolous – n. Inhabiting limestone soils.

Gypsophilous–a. Limestone-loving; inhabiting gypsum soils.

Loam

Melangeophilous – n. Loam-loving; inhabiting loam soils.

Marshes

Banados – pl. n. Shallow swamps (Paraguay).

Helad – n. A marsh plant.

Limnodophilous – a. Dwelling in marshes.

Paludose – a. Growing in marshy places.

Palustrine – a. Of or growing in marshes.

Pontohalicolous – a. Dwelling in salt marshes.

Stasad – n. A plant of stagnant water.

Telmatophilous – a. Loving wet meadows.

Telmicolous – a. Dwelling in fresh water marshes.

Mountains

Montane – a. Pertaining to mountains, as plants which grow on them.

Orophilous – a. Mountain-loving; inhabiting mountains.

Mud

Limicolous – a. Inhabiting muddy places, as on the margins of pools.

Limose – a. Of marshes.

Luticolous – a. Mud- or mire-loving; inhabiting muddy places.

Prairie

Graminicolous – a. Grass-inhabiting.

Psilicolous – a. Prairie-dwelling.

Psilophilous – a. Prairie-loving.

Rain

Ombrophilous – a. Rain-loving, inhabiting places of frequent rains.

Salt

Drimophilous – a. Salt-loving.

Halophilous – a. Salt-loving; growing in salty soils.

Halophyte – n. A plant which grows in saline soil.

Saline – a. Of or pertaining to salt.

Sand

Amathicolous – a. Growing on sandy plains.

Amathophilous – a. Growing in sandy plains or in sandy hills.

Ammochthad – n. A sand-bank plant.

Ammophilous – a. Sand-loving; inhabiting sand.

Arenarious – a. Of sand or sandy places.

Arenicolous – a. Inhabiting sandy places.

Cheradad – n. A sand-bar plant.

Cheradophilous – a. Loving dry habitats; dwelling in dry places.

Enaulophilous – a. Loving sand draws.

Psammophilous – a. Sand-loving; inhabiting sand.

Sabulicolous – a. Growing in sandy places.

Thinicolous–a. Dwelling on shifting sand dunes.

Thinophilous – a. Dune-loving; inhabiting dunes.

Sea

Agad – n. A beach plant.

Aigialophilous – a. Beach dwelling.

Aigicolous – a. Inhabiting a stony strand or beach.

Littoral – n. Belonging to or growing on the seashore.

Marine – a. Growing within the influence of the sea, or immersed in it.

Maritime – a. Belonging to the sea, or confined to the sea coast.

Shade

Sciophilous – a. Shade-loving; inhabiting shady places.

Umbracticolous – a. Inhabiting shady places.

Snow

Chionad – n. A snow plant.

Chionic – a. Of snow fields.

Niveous – a. Growing in or near the snow, pertaining to snow.

Springs

Crenad – n. A plant of springs.

Crenophilous – a. Loving springs.

Stone

Chasmophilous – a. A cranny-loving plant.

Lapidose – a. Growing among stones.

Petraeous – a. Of or pertaining to stones.

Petricolous – a. Rock-inhabiting.

Petrophilous – a. Stone-loving, dwelling among stones.

Phellophilous – a. Loving rock fields.

Rupestral – a. Pertaining to rocks.

Silicolous – a. Growing in flinty soils.

Streams

Crenicolous – a. Dwelling in spring-fed brooks.

Fluvial – a. Said of plants growing in streams.

Namatad – n. A plant growing in or near a brook.

Potomophilous – a. River-loving; dwelling in or near rivers.

Rheophilous – a. Creek-loving; dwelling in torrents.

Rhyacophilous – a. Torrent-loving; dwelling in torrents.

Riparious – a. Growing by rivers or streams.

Sun

Heliad – n. A heliphyte or sun-loving plant.

Thickets

Aithalophilous – a. Dwelling in evergreen thickets.

Capoe – n. A palm thicket (Brazil).

Driodad – n. A plant of a dry thicket.

Lochmocolous – a. Inhabiting thickets.

Lochmodophilous – a. Loving dry thickets; growing in dry thickets.

Lochmophilous – a. Thicket-loving; found in dry thickets.

Walls

Rupestral – a. Growing on walls and rocks.

Waste places

Chledocolous – a. Dwelling in waste places.

Chledophilous – a. Loving waste places.

Water

Emersed – a. Raised above and and out of the water, emerged.

Hydrophilous – a. Loving wet places or water; pollinated by water.

Hydrophyte – n. A water plant.

Natant – a. Floating.

Inflorescence

Ament – n. A catkin; a more or less flexible, usually pendulous spike bearing apetalous unisexual flowers.

Amentiferous – a. Bearing aments.

Amentum – n. Catkin.

Anthela – n. The panicle of *Juncus* with the lateral axes exceeding the main axis.

Anthelate–a. With elongate flower-bearing branches, as in some *Junci*.

Anthemy, anthemia – n. A flower-cluster of any kind.

Capitate – n. With a head.

Catkin – n. See Ament.

Centrifugal – a. In inflorescences, blooming from the inside outward, or from the top downward.

Centripetal – a. In inflorescences, blooming from the outside inward, or from the base upward.

Cincinnus–n. A one-branched scorpioid cyme.

Compound inflorescence – n. An inflorescence composed of secondary inflorescences.

Corymb – n. A short, broad, more or less flat-topped, indeterminate flower-cluster, the outer flowers opening first.

Corymbiform – a. In the shape of a corymb.

Corymbose – a. Arranged in a corymb.

Cyanthum – n. The ultimate inflorescence of *Euphorbia* consisting of a cuplike involucre bearing the flowers from its base.

Cyme – n. A broad, more or less flat-topped, determinate flower-cluster, with central flowers blooming first.

Cymose – a. Cymelike.

Cymule – n. A small cyme.

Definite – a. Determinate, terminating in a flower bud.

Definite inflorescence – n. A determinate inflorescence, terminating in a flower bud, blooming from the inside outward or from the top downward.

Determinate – a. Said of an inflorescence in which the terminal flower blooms slightly in advance of its nearest associates.

Dichasium – n. A cyme with two lateral axes.

Dicymose – a. Doubly cymose.

Diffuse – a. Loosely branching or spreading; of open growth.

Drepanium – n. A sickle-shaped cyme.

Ecblastesis – n. The appearance of buds within a flower, proliferation of an inflorescence.

Fascicle – n. A condensed or close cluster.

Fasciculate – a. In condensed or close clusters.

Glome – n. A rounded head of flowers.

Glomerate – a. In a dense or compact cluster or clusters.

Glomerule – n. A cluster of heads in a common involucre.

Head – n. A dense spherical or flat-topped inflorescence of sessile flowers clustered on a common receptacle.

Indefinite – a. In an inflorescence, indeterminate.

Indefinite inflorescence – n. An inflorescence that is indeterminate, blooming from the outside inward, or from the bottom upward.

Indeterminate – a. Descriptive of an inflorescence in which the flowers open progressively from the base upward or from the outside inward.

Inflorescence – n. Mode of flower-bearing; technically less correct but much more common in the sense of a flower-cluster.

Intercalary inflorescence – n. An inflorescence in which the main axis continues to grow vegetatively after giving rise to the flowers.

Julaceous – a. Bearing catkins, amentaceous.

Mixed inflorescence – n. One in which partial inflorescences develop differently from the main axis, as centrifugal and centripetal together.

Monochasium – n. A one-branched cyme, either pure or resulting from the reduction of cymes.

Nucamentum – n. An amentum, or catkin.

Panicle – n. A compound or branched raceme.

Paniculate – a. Having a panicle type of inflorescence.

Phoranthium – n. The receptacle of the head of *Compositae*.

Polythalamic – a. Having more than one female flower within the involucre; derived from more than one flower, as a collective fruit.

Raceme – n. An indeterminate inflorescence consisting of a central axis bearing a number of flowers with pedicels of nearly equal length.

Racemose – a. Resembling a raceme; in racemes.

Racemiform – a. In the form of a raceme.

Rhipidium – n. A fan-shaped cyme, the lateral branches being developed alternately in two opposite directions.

Scape – n. A leafless peduncle arising from the ground; it may bear scales or bracts but not foliage-leaves and may be one- or many-flowered.

Scapiform – a. Resembling a scape.

Scapose – a. Bearing or resembling a scape.

Scorpioid – a. Said of a coiled cluster in which the flowers are two-ranked and borne alternately at the right and left.

Scorpioid cyme – n. Cincinnus, the lateral branches developed on opposite sides alternately, as in *Boraginaceae*.

Simple inflorescence – n. A flower-cluster with one axis, as a spadix, spike, or catkin.

Spadix – n. The thick or fleshy spike of certain plants, as the *Araceae*, surrounded or subtended by a spathe.

Spathe – n. The bracts or leaf surrounding or subtending a flower-cluster or spadix; it is sometimes colored and flower-like, as in the *Calla*.

Spiciform – a. Spikelike.

Spike – n. An inflorescence consisting of a central rachis bearing a number of sessile flowers.

Tassel – n. The staminate inflorescence in maize.

Thyrse – n. Compact and more or less compound panicle; more correctly a panicle-like cluster with main axis indeterminate and other parts determinate.

Thyrsoid – a. Resembling a thyrse.

Thyrsula – n. A little cyme which is borne by most labiates in the axil of the leaves.

Trichasium – n. A cymose inflorescence with three branches.

Umbel – n. An indeterminate inflorescence consisting of several pedicellate flowers having a common point of attachment.

Umbel, compound – n. An umbel in which each ray bears a small umbel.

Umbel, cymose – n. An apparent umbel, but with the flowers opening centrifugally; a cyme which simulates an umbel.

Umbellate – a. With or pertaining to umbels.

Umbellet – n. A secondary umbel.

Umbelliferous – a. With umbels.

Umbelliform – a. In the shape of an umbel.

Umbellule–n. An umbellet; a small umbel.

Verticillaster – n. A false whorl, composed of a pair of opposed cymes, as in labiates.

Leaves

Forms

Aciculate – a. Slender, needle-shaped.

Awl-shaped – a. Narrow and gradually tapering to a sharp point.

Cochlear – a. Said of a form of imbricate aestivation with one piece exterior.

Cordate – a. Heart-shaped, with the notch basal.

Cuneate – a. Wedge-shaped, with the broad end apical.

Deltoid – a. Triangular; delta-shaped.

Drepaniform – a. Sickle-shaped.

Elliptical – a. Oblong with rounded ends.

Ensiform – a. Sword-shaped, gladiate.

Falcate – a. Sickle- or scythe-shaped.

Gladiate – a. Sword-shaped, ensiform.

Halberd-shaped – a. Sagittate with the basal lobes turned outward, hastate.

Hastate – a. Arrow-shaped with the basal lobes turned outward, halberd-shaped.

Heart-shaped – a. Cordate, broadly ovate with two rounded lobes at the base.

Lanceolate–a. Lance-shaped, much longer than wide and tapering upward.

Linear – a. Long and narrow with margins parallel or nearly so.

Lyrate – a. Lyre-shaped.

Needle – n. The stiff linear leaf of a *Pinaceae*.

Nephroid – a. Kidney-shaped, reniform.

Obcordate – a. Heart-shaped with the notch apical.

Oblanceolate – a. Inverted lanceolate.

Oblong – a. Longer than broad, with the margins nearly parallel.

Obovate – a. Oval, but broader toward the apex.

Orbiculate – a. Round or circular.

Oval – a. Elliptical with the width greater than half the length.

Ovate – a. Oval, but broader toward the base.

Pandurate – a. Fiddle-shaped.

Panduriform – a. Fiddle-shaped.

Peltate – a. Shield-shaped with the petiole attached to the under side.

Reniform – a. Kidney-shaped nephroid.

Rhomboidal – a. Rhombic-shaped.

Sagittate– a. Arrow-shaped.

Scale – n. Any thin scarious body, usually a degenerate leaf, sometimes of epidermal origin; sometimes used meaning glume.

Scale-leaves – n. Modified leaves on underground stems; small flat leaves as those on *Cupressus* and *Selaginella*.

Spatulate - a. Spatula-shaped.
Subulate - a. Awl-shaped.
Wedge-shaped - a. Cuneate.

Apexes

Acuminate - a. Tapering to a prolonged point.
Acute - a. Distinctly and sharply pointed, but not drawn out.
Apiculate-a. With a minute pointed tip.
Aristate - a. Awned, bearing an arista.
Cuspidate - a. Tipped with a sharp rigid point.
Emarginate-a. With a shallow notch at the apex.
Mucronate - a. With a mucro; bristle-tipped.
Obcordate - a. Heart-shaped with the notch at the apex.
Obtuse - a. Blunt or rounded at the end.
Retuse - a. An obtuse tip with a slight depression in the middle.
Truncate - a. As though cut off by a straight transverse line.

Bases

Acuminate - a. With prolonged tapering to the petiole.
Acute - a. Distinctly and sharply pointed but not drawn out.
Amplexicaul - a. Clasping the stem.
Auriculate - a. With an auricle or a claw.
Clasping - a. With the base clasping the stem.
Connate-perfoliate - a. Having opposite leaves joined at the bases.
Cordate - a. Heart-shaped with the notch at the base.
Cuneate - a. Wedge-shaped.
Decurrent - a. Said of a leaf which extends down a stem below the point of insertion.
Hastate - a. Similar to sagittate but with the lobes pointing outward.
Oblique - a. With one side of the base being larger than the other.
Obtuse - a. Blunt or rounded at the end.
Ocrea - n. A legging-shaped or tubular structure formed by the union of two stipules.
Peltate - a. With the petiole attached to the under side rather than the margin.
Perfoliate - a. With the stem apparently passing through the leaf.
Sagittate-a. With basal lobes pointing downward, like the base of an arrow.
Surcurrent - a. With winged expansions from the base of a leaf prolonged up the stem.
Truncate - a. As if cut off by a straight transverse line, blunt.

Modifications

Bracts	Scales
Fronds	Sepals
Petals	Spines
Phyllode	Stamens
Pistils	Tendrils

Complexity

Simple - a. With one blade with incomplete or no segmentation.
Compound - a. With two or more blades called leaflets.
- **Pinnately** - adv. With the leaflets arranged on opposite sides along a common rachis.
  - **Odd pinnate** - a. With a terminal leaflet.
  - **Abrupt pinnate** - a. Without a terminal leaflet.
- **Palmately** - adv. With the leaflets arising from the apex of the petiole in a palmate manner.
- **Radiately** - adv. With the leaflets radiating in all directions from the apex of the petiole.

Decompound - a. More than once compound.

Margins

Cleft - a. Divided into lobes separated by narrow or acute sinuses which extend more than half way to the midrib.
Crenate - a. With rounded or blunt teeth.

Crenulate – a. Finely crenate.
Crisp – a. Curly or wavy, as the leaves of *Rumex crispus*.
Dentate–a. With sharp teeth pointing outward.
Denticulate – a. Minutely or finely dentate.
Dissected – a. Deeply divided or cut into many segments.
Divided – a. With lobing or segmentation extending to the base or midrib.
Doubly serrate – a. With small serrations on larger ones.
Entire – a. With an even margin; not interrupted by toothing, lobing, or other divisions.
Fimbriate – a. With the margin bordered with long slender processes.
Glandulose-serrate – a. With serrations tipped or bordered with glands.
Incised – a. Cut sharply and irregularly, more or less deeply.
Inflexed – a. Turned in at the margins.
Intramarginal – a. Within and near the margins.
Lacerated – a. Torn or irregularly cleft.
Laciniate – a. Cut into lobes separated by deep, narrow, irregular incisions.
Lobed – a. With lobes extending to near the middle.
Multifid – a. Cleft into many lobes or segments.
Palmate – a. Lobed or divided, so that the sinuses point to the apex of the petiole.
Palmatifid – a. Lobed or divided, so that the sinuses point nearly to the apex of the petiole.
Parted – a. Divided by sinuses which extend nearly to the midrib.
Partitioned – a. Having the deepest division into which a leaf can be cut without becoming compound.
Pectinate – a. With narrow segments set close together like the teeth of a comb.
Peltate–a. With the petiole attached to the under side instead of the margin.
Peltified – a. Said of a peltate leaf that is cut into segments.
Peripheral – a. On or near the margins.
Pinnatifid – a. Cleft almost to the midrib.
Repand – a. Undulate or wavy.
Revolute – a. With margins rolled toward the lower side.
Runcinate – a. Saw-toothed or sharply incised with retrorse teeth.
Serrate – a. With sharp teeth pointing forward.
Serrulate – Serrate with minute teeth.
Sinuate – a. With a deep wavy margin.
Sinuous – a. Wavy.
Undulate – a. Wavy, repand.

Venation

Anadromous – a. Said of ferns, in which the first set of nerves in each segment of the frond is given off on the upper side of the midrib toward the apex, as in *Aspidium* and *Asplenium*.
Anametadromous – a. Said of the venation of ferns in which the weaker pinnules are anadromous and the stronger are catadromous.
Anastomosing – a. Characterized by the union of one vein with another, the connection forming a reticulation.
Arcuate–a. Moderately curved, bent like a bow, said of leaf venation of *Cornus, Caenothus*, etc.
Argyroneurous – a. With silver-colored veins.
Basinerved – a. Veined from the base.

Campylodromous – a. Said of venation in which the secondary veins curve toward the margins but do not form loops.

Catadromous – a. Said of the venation of ferns in which the first set of nerves in each segment of the frond is given off on the basal side of the midrib, as in *Osmunda.*

Convergent – a. Said of veins which run from the base to the apex of a leaf in a curved manner.

Costa – n. A rib, as a midrib.

Craspedodromous – a. With the lateral veins running from midrib to margin without dividing.

Diadromous – a. With fan-shaped venation, as in *Gingko biloba.*

Dictyodromous – a. With reticulate venation.

Feather-veined – a. With secondary veins proceeding from the midrib; pinninerved.

Hinoideous – a. With veins proceeding from the midrib parallel and unbranched.

Hyphodromous – a. With the veins sunken in the leaf and not readily visible.

Infossous–a. With the veins sunken but leaving a visible channel.

Intercostal – a. Between the ribs or veins.

Internerves – n. The space between the nerves.

Leuconeurous – a. White-nerved.

Marmorate – a. With veins of color, as marbled, or mottled.

Midrib – n. The main rib or central vein of a leaf or leaflike structure.

Multicostate – a. Many-ribbed.

Nervation – n. Venation, the manner in which foliar nerves or veins are arranged.

Nerve – n. A simple or unbranched vein or slender rib.

Net-veined – a. Reticulated; net-veined with any system or irregularly anastomosing veins.

Palmately veined – a. With veins arranged in a palmate manner.

Parallelodromous – a. With parallel veins, as in lilies.

Parallel-veined – a. With the lateral veins straight, as in *Alnus;* with the entire system straight as in grasses.

Penniveined – a. Veined in a pinnate manner.

Pinninerved – a. Pinnately veined, the veins running parallel towards the margin.

Plagiodromous – a. Said of tertiary leaf-veins when at right angles to the secondary veins.

Quinquenerved – a. With the midrib dividing into five, that is, the main rib and a pair on each side.

Radiately veined – a. With veins radiating from a centrally attached petiole.

Reticulate – a. Forming a network.

Rib – n. A primary vein, especially the central longitudinal or midrib.

Trinervate – a. Three-nerved.

Vein – n. A strand of vascular tissues in a flat organ, such as a leaf.

Venation – n. Veining; the arrangement or disposition of veins.

–onyms

Antonym – n. A word of opposite meaning; a counterterm.

Basonym – n. The original epithet, retained when transferred to a new position.

Homonym – n. A name having the same spelling as another name in the same language but different in meaning; in taxonomy, homonyms are two or more names having the same spelling but applied to two or more taxa of the same rank based upon different types. A later homonym is illegitimate.

Hyponym – n. A name to be rejected for want of an identified type.

Metonym – n. A name that is rejected because there is an older valid name based on another member of the same group.

Synonym – n. A name with the same meaning as another name in the same language but spelled differently; in taxonomy, synonyms are two or more scientific names for the same taxon, one of which is correct and the others incorrect under the International Rules of Nomenclature.

Synonymous – a. Having the same meaning.

Synonymy – n. Discarded names for identical objects.

Typonym – n. A name rejected because an older name was based on the same type.

Ovules and Seeds

Albuminous – a. Having albumin or an endosperm.

Amphitropous – a. Half-inverted and straight, with the hilum lateral.

Anatropous – a. With the ovule reversed, with the micropyle close to the side of the hilum and the chalaza at the opposite end.

Angiospermous – a. With the seeds borne within a pericarp.

Angiosperm – n. A plant having its seeds enclosed in an ovary.

Aril – n. An appendage or an outer covering of a seed, growing out from the hilum or funiculus; sometimes it appears as a pulpy covering.

Campylotropous – a. Said of an ovule or seed which is curved in its formation so as to bring the micropyle or true apex down near the hilum.

Caruncle – n. An excrescence or appendage at or about the hilum of the seed.

Chalaza – n. That part of the ovule or seed where the nucellus joins the integuments; it is the base of the nucellus and is always opposite the upper end of the cotyledons.

Cotyledon – n. A seed leaf; a primary leaf in the embryo.

Dicotyledones – pl. n. Plants of the class denoted by their possession of two cotyledons.

Dispermous – a. Two-seeded.

Embryo – n. The rudimentary plant formed in the seed.

Endosperm – n. The albumen of a seed in angiosperms; in gymnosperms, the prothallium within the embryo sac.

Exalbuminous–a. Without albumin, used only of seed in which the embryo occupies the entire cavity within the testa.

Foramen – n. An aperture, especially that in the outer integuments of the ovule; micropyle.

Funicle – n. See Funiculus.

Funiculus – n. A stem or thread-like structure that connects the ovule or seed to the placenta.

Hilum – n. The scar or mark on a seed indicating the point of attachment.

Hylum – n. See Hilum.

Hypocotyl – n. The axis of an embryo below the attachment of the cotyledons.

Incumbent cotyledons – pl. n. Cotyledons so arranged that the back of one lies against the radicle.

Integument – n. The envelope of an ovule; the seed coat.

Kernel – n. The nucellus of an ovule, or of a seed, that is, the whole body within the seed coats.

Melanospermous – a. With black or dark-colored seeds, or spores.

Mesocotyl – n. An interpolated node in the seedling of grasses, so that the sheath and cotyledon are separated by it.

Micropyle – n. The aperture in the integument of a seed formerly the foramen; it marks the position of the radicle.

Monocotyledon – n. A plant having but one cotyledon or seed lobe.

Monocotyledonous – a. With only one cotyledon.

Monospermous – a. One-seeded.

Nucleus – n. A kernel of an ovule, which by fertilization becomes a seed; a dense protoplasmic structure near the center of living cells.

Oligospermous – a. Few-seeded.

Oncospore – n. A seed with hooks which aid in dispersal.

Oöspore – n. The fertilized egg in the archegonium of cryptogams from which the new plant develops directly.

Orthotropous – a. Said of an ovule or seed with a straight axis, chalaza at the insertion, the orifice at the other end.

Ovuliferous – a. Bearing ovules.

Parthenogenesis – n. The production of seeds without fertilization.

Perisperm – n. The ordinary albumen of a seed, restricted to that which is formed outside the embryo sac; the integuments of the seed.

Pip – n. A popular name for the seed of an apple or pear.

Pleiospermous – a. With an unusually large number of seeds.

Plumule – n. The primary leaf-bud of an embryo.

Primine – n. The outer integument of an ovule.

Pterospermous – a. With winged seeds.

Raphe – n. An adnate cord or ridge or fibro-vascular tissue which, in a more or less anatropous ovule, connects the hilum with the chalaza.

Rhaphe – n. See Raphe.

Seed – n. A mature ovule.

Seed leaf – n. A cotyledon.

Semen – n. The seed of flowering plants.

Seminiferous – a. Seed-bearing; used for the special portion of the pericarp bearing the seeds.

Strophiole – n. An appendage at the hilum of certain seeds; a caruncle.

Testa – n. The outer coat of the seed, usually hard and brittle.

Umbilical cord – n. A vascular strand by which seeds are attached to the placenta; the funiculus.

Umbilicus – n. The hilum of the seed.

Pistils

Adynamogyny – n. Loss of function in the female organ of a flower.

Agamogynaecism – n. In *Compositae,* the state of having female and neuter flowers in the same individual.

Apocarpous – a. With separate carpels, not syncarpous.

Apocarpy – n. The condition of having the carpels separate.

Bursicule – n. The pouchlike expansion of the stigma into which the caudicle of some orchids is inserted.

Carpel – n. A simple pistil; one unit of a compound pistil; in conifers, the cone scale of the female cone.

Carpellate – a. Possessing carpels.

Compound pistil – n. With two or more carpels coalesced into one body.

Dodecagynous–a. Possessing twelve distinct pistils or carpels.

Gynoecium – n. The pistil or pistils of a flower; the female portion as a whole.

Gynophore – n. The stipe or stalk of an ovary prolonged within the calyx.

Gynosporangium – n. The receptacle in which gynospores are developed.

Gynospore – n. One of the larger reproductive bodies (female) in the *Isoëtaceae.*

Heterodistyly – n. Dimorphism, the presence of two kinds of plants within a species, one with long, the other with short styles.

Hexagynous – a. With six pistils.

Homostyly – n. The same relation of length between all styles and anthers of the same species.

Hypogynium – n. The perianth-like structure subtending the ovary in *Scleria* and some other *Cyperaceae.*

Hypogynous – a. Free from, but inserted beneath, the pistil or gynoecium.

Inferior – a. Beneath, lower, as an inferior ovary, one that is below the calyx-leaves.

Inferior ovary – n. An ovary with the perianth located on top.

Macrostylous – a. Long-styled.

Mesocarp – n. The middle layer of a pericarp.

Metandry – n. A condition in which the female flowers mature before the male; protogyny.

Monogynous – a. With one carpel.

Monostylous – a. With a single style.

Octagynia – n. A Linnean order of plants with eight-styled flowers.

Octagynous – a. With eight-styled flowers.

Ovary – n. That part of the pistil which contains the ovules.

Pentagynous – a. With five pistils or styles.

Perigynium – n. The hypogynous setae of sedges; the flask or utricle of *Carex*; any hypogynous disc.

Perigynous – a. Borne around the ovary and not beneath it, as with calyx, corolla, and stamens borne on the edge of a cup-shaped hypanthium.

Pistil – n. The female organ of a flower, consisting when complete of an ovary, style, and stigma.

Pistillate – a. With pistils and no stamens; female.

Protogynous – a. With pistil maturing before stamens in the same flower.

Pterogynous – a. With a winged ovary.

Scutum – n. The broad dilated apex of the style in asclepiads.

Simple pistil – n. A pistil of one carpel; not compound.

Stegium – n. Threadlike appendages sometimes found covering the style of asclepiads.

Sterile – a. Barren, as a flower without a pistil.

Stigma – n. The part of a pistil that receives the pollen; a point on the spores of *Equisetum.*

Stigmatic – a. Pertaining to the stigma.

Stylar – a. Relating to the style.

Style – n. The more or less elongated part of the pistil between the ovary and the stigma.

Stylopod – n. The enlargement at the base of the styles in *Umbelliferae.*

Superior ovary – n. An ovary with all of the floral envelopes inserted below on the torus.

Positions

Accumbent – a. Lying against another organ.

Accumbent cotyledons–pl. n. Cotyledons having their edges against the radicle.

Appressed – a. Lying flat against an organ.

Ascending – a. Sloping upward; produced somewhat obliquely or indirectly upward.

Assurgent – a. Ascending, rising.

Cernuous – a. Drooping; inclining somewhat from the perpendicular, nodding.

Coarctate – a. Crowded together.
Decumbent – a. Reclining or lying on the ground, but with the ends ascending.
Descending – a. Tending gradually downward, as the branches of some trees and as the roots.
Dextrorse – a. Turning to the right, or clockwise.
Erect – a. Upright, perpendicular.
Extrorse – a. Turned or faced outward or away from the axis, as an anther turned away from the center of the flower.
Incumbent – a. Resting or leaning upon.
Introrse – a. Turned or faced inward or toward the axis, as an anther turned toward the center of the flower.
Inverted – a. Turned over; end-for-end; top side down.
Juxtaposition – n. The relative position in which organs are placed; a placing or being placed side by side.
Nodding–a. Curved somewhat from the vertical, drooping.
Procumbent – a. Lying upon the ground, prostrate, trailing.
Prostrate – a. Lying flat, procumbent.
Reclinate – a. Bent down or falling back from the perpendicular.
Recurved – a. Bent or curved downward or backward.
Reflexed – a. Abruptly curved or bent downward or backward.
Repent – a. Prostrate and rooting.
Reptant–a. Creeping on the ground and rooting.
Resupinate – a. Upside down, or apparently so.
Retrocurved – a. Bent or curved back.
Retroflexed – a. Bent back, reflexed.
Retrorse – a. Directed backwards or downwards.
Scandent – a. Climbing in any manner.
Sinistrorse – a. Turning to the left, or counterclockwise.
Strict – a. Stiff, upright, rigid.
Subterraneous – a. In or under the soil.
Supine – a. Prostrate with face turned up.
Trailing–a. Prostrate, but not rooting.

Sepals

Asepalous – a. Without sepals.
Caducous – a. Falling off early or prematurely, as the sepals in some plants.
Calyanthemy – n. Petalody of the calyx; the formation of colored petal-like structures in place of a normal calyx.
Calyculate – a. Calyx-like, bearing a part resembling a calyx, particularly, with bracts against or underneath the calyx resembling a supplementary or outer calyx.
Calyx – n. The outermost circle of floral envelopes.
Disepalous – a. With two sepals.
Epicalyx – n. A series of bracts close to and resembling the calyx.
Gamosepalous – a. With the sepals united.
Monosepalous – a. With one sepal.
Octosepalous – a. With eight sepals.
Pentasepalous – a. With five sepals.
Pleiosepalous – a. Many-sepaled.
Polysepalous – a. With many distinct sepals.
Sepal – n. One of the separate parts of a calyx.

Sex Distribution

Agamohermaphrodite – a. With hermaphrodite and neuter flowers on the same plant.
Agamospermy – n. Seed production without fertilization.
Allautogamia – n. The state of having two methods of pollination, one usual, and the other facultative.

Allogamous – a. Reproducing by cross-fertilization.
Allogamy – n. The pollination of a flower with pollen from another flower; see **Geitonogamy** and Xenogamy.
Amixia – n. Cross-sterility.
Androgynous – a. Hermaphrodite; with both male and female flowers in the same inflorescence; occasionally used meaning monoecious.
Autoicous – a. See Monoecious.
Autophilous – a. Self-pollinated.
Cenanthy – n. The suppression of the stamens and pistil leaving the perianth empty.
Column – n. A combination of stamens and styles into a solid central body, as in orchids.
Cleistogene – n. A plant which bears cleistogamous flowers.
Cleistogeny – n. The state of bearing cleistogamous flowers.
Cleistogenous – a. Cleistogamous.
Dichogamous – a. Hermaphrodite with one sex maturing earlier than the other, the stamens and pistil not synchronizing.
Dichogamy – n. A condition in perfect flowers in which the sexes do not mature simultaneously.
Diclinism – n. The separation of anthers and stigma in space, as dichogamy does in time.
Diclinous – a. Unisexual, having the stamens in one flower and the pistil in another.
Diecious – a. See Dioecious.
Dimorphic – a. Occurring under two forms, as with long and short styles.
Dioecious – a. Unisexual, with staminate flowers on one plant and the pistillate on another.
Disanthic – a. With fertilization by pollen from another plant.
Epigymous – a. Borne on the ovary, the ovary inferior and not perigynous.
Exogynous–a. With the style longer than the corolla and projecting beyond it.
Flos. (pl. **flores**) – n. A perfect flower with some protecting envelope.
Frustraneous – a. Said of *Compositae* with disk flowers hermaphrodite, and those of the ray neuter or imperfect.
Geitonogamy – n. Pollination by pollen from another flower on the same plant.
Gnesiogamy – n. Fertilization between different individuals of the same species.
Gynandrous – a. With the stamens adnate to the pistil, as in orchids.
Gynecandrous – a. With staminate and pistillate flowers in the same spike, the pistillate at the apex; used chiefly in the *Cyperaceae*.
Gynodioecious – a. Dioecious with some flowers hermaphrodite, others pistillate only, on separate plants.
Gynomonoecious – a. Monoecious with female and hermaphrodite flowers on the same plants.
Gynostemium – n. The compound structure resulting from the union of the stamens and pistil in *Orchidaceae*.
Hermaphrodite – a. With stamens and pistils in the same flower.
Heteracmy – n. The condition in which stamens and pistils mature at different times.
Heterogamous–a. Bearing two kinds of flowers.
Homocephalic – a. Delpino's term in reference to homogamy in which the pollen fertilizes another flower in the same inflorescence.
Heterocephalous – a. With staminate and pistillate flowers on separate heads on the same plant.
Homoclinous – a. Delpino's term in reference to that kind of homog-

amy in which a complete flower is fertilized by its own pollen.

Homogamous - a. Bearing one kind of flower.

Homogony - n. The condition in which the pistils and stamens of all flowers are of uniform relative length.

Homotropic - a. Fertilized by pollen from the same flower.

Imperfect - a. Said of flowers lacking one of the essential organs.

Misogamy - n. Reproductive isolation.

Monoclinous - a. Having both stamens and pistils in the same flower; applied to the heads of *Compositae* that have only hermaphrodite flowers.

Monoecious - a. Having unisexual flowers with both sexes on the same plant.

Motion-dichogamy - n. A condition in which the sexual organs vary in length or position during flowering.

Neuter - a. Sexless, as a flower that has neither stamens nor pistils.

Neutral - a. Without stamens or pistils, sexless.

Nyctigamous - a. Said of flowers which open at night and close by day, marrying at night.

Panmixy - n. Free and more or less unlimited cross-fertilization.

Perfect - a. Said of flowers having both sex organs present and functioning.

Phaenogamous - n. Said of plants sexually propagating by flowers, the essential organs of which are stamens and pistil.

Phenological isolation - n. Isolation by a time of flowering, as either earlier or later than the other species of the genus.

Phytogamy - n. Cross-fertilization of flowers.

Polygamodioecious-a. Polygamous, but chiefly dioecious.

Polygamomonoecious - a. Polygamous, but chiefly monoecious.

Polygamous - a. Bearing perfect and unisexual flowers on the same individual.

Psychrocleistogamy-n. Cleistogamy induced by low temperature.

Staurigamia - n. Cross-fertilization.

Superior - a. Growing or placed above; hypogynous.

Synacmy - n. A condition in which stamens and pistil mature simultaneously; the opposite of heteracmy.

Synanthesis - n. The simultaneous maturation of stamens and pistil; synacmy.

Synoecious - a. With staminate and pistillate flowers both present in the same head.

Thermocleistogamy - n. Self-pollination taking place within flowers the opening of whose perianth has been inhibited by low temperature.

Trigamous - a. Bearing three forms of flowers; trimorphic.

Trimorphic - a. Occurring in three forms, as with long, short, and intermediate styles.

Unisexual - a. Of one sex; stamens or pistil only.

Xenogamy - n. Cross-fertilization between sexual elements borne by different individuals.

Stamens

Agamandroecism - n. The condition in *Compositae* of having male and neuter flowers in the same individual.

Agynic - a. Said of stamens which are free from the ovary; destitute of pistils.

Androecium - n. The stamens of a flower (a collective term).

Androphore - n. A support or column on which the stamens are raised.

Anther – a. That portion of the stamen which bears the pollen.

Antheriferous – a. Anther-bearing.

Antheroid – a. Anther-like.

Bicruris – a. Two-legged, as the pollen masses of asclepiads.

Clinandrium – n. The anther bed in orchids, that part of the column in which the anther is concealed.

Cryptandrous – n. With hidden anthers, clesistogamous, the stamens remaining enclosed in the flower.

Decandrous – a. With ten stamens.

Diadelphous – a. With stamens formed in two groups by the union of their filaments.

Diandrous – a. With two stamens.

Didynamous – a. Four-stamened with stamens in pairs, two long, two short, as in many labiates.

Dodecandrous – a. Normally possessing twelve stamens, occasionally extended to more than twelve.

Emasculation – n. The removal of the anthers from a bud or flower.

Enneandrous – a. With nine stamens.

Exserted – a. Sticking out, projecting beyond, as stamens from perianth; not included.

Filament – n. The part of a stamen that supports the anther.

Gynostegium – n. A sheath or covering of the gynoecium, of whatever nature.

Heterandrous – a. With two sets of stamens, as flowers with two kinds of stamens.

Hexandrous – a. With six stamens.

Homoeandrous – a. With only one kind of stamen.

Icosandrous – a. With twenty or more stamens.

Incumbent anther – n. An anther attached to the inner face of its filament.

Isadelphous – a. Equal brotherhood, the number of stamens in each group being equal.

Massule – n. A group of cohering pollen-grains produced by one primary mother cell in orchids, also styled pollen-mass.

Melantherous – a. With black anthers.

Monadelphous – a. With stamens united by their filaments into a tube or column.

Monandrous – a. With one stamen.

Nototribal – a. With stamens arranged so as to deposit pollen on the backs of their insect visitors.

Octostemonous – a. With eight fertile stamens.

Oligandrous – a. With few stamens.

Parastemon – n. An abortive stamen, a staminode.

Pentadelphous – a. With five fraternities or bundles of stamens.

Pentandrous – a. With five stamens.

Phaenantherous – a. With stamens exserted.

Pleurotribal – a. Said of flowers whose stamens are adapted to deposit their pollen upon the sides of insect visitors.

Polyadelphous – a. With stamens disposed into several brotherhoods or groups.

Polyandrous – a. With an indefinite number of stamens.

Polystemonous – a. Polyandrous; with numerous stamens.

Porandrous–a. With anthers opening by pores.

Protandrous – a. With the anthers maturing before the pistil in the same flower, one kind of dichogamy.

Protandry–n. A condition in which the stamens mature before the pistil in the same flower.

Proterandrous – a. With the stamens maturing before the pistil in the same flower; protandrous, one kind of dichogamy.

Proterogyny - n. A condition in which the pistil matures before the anthers.

Psilostemon - a. Smooth-stamened; naked-stamened.

Resilient - a. Springing or bending back, as some stamens.

Sporophyll - n. A spore-bearing leaf.

Stamen - n. The pollen-bearing organ of the flower, the male organ.

Stamen, sterile- n. A body belonging to the series of stamens but without pollen.

Staminate - a. With stamens but no pistil; male.

Stamineal - a. Relating to or consisting of stamens.

Staminode - n. A sterile stamen, or a structure resmbling such and borne in the staminal part of the flower; in some flowers (as in *Canna*) staminodia are petal-like and showy.

Sternotribal - a. Said of flowers whose anthers are so arranged as to dust their pollen on the under part of the thorax of their insect visitors.

Sulphur rain - n. Pine pollen carried in excessive amounts by air currents.

Synantherous - a. With anthers joined to form a tube.

Synema (pl. **synemata**) - n. The column of monadelphous stamens, as in *Malvaceae*.

Triadelphous - a. With stamens in three sets.

Triandrous-a. With three stamens.

Tridynamous - a. With three stamens out of six being longer than the rest.

Stems

Acaulescent - a. Stemless or apparently so.

Armed - a. Possessing any kind of strong and sharp defense, as thorns, spines, prickles, or barbs.

Articulate - a. Jointed; with nodes or joints, or places where separation may naturally take place.

Axil - n. The upper angle formed between the axis and any organ that arises from it.

Axillary - a. Situated in the axil.

Axis - n. The main or central line of development of any plant or organ; the main stem.

Bast - n. The fibrous constituent of the bark of many species.

Bole - n. The main trunk of a tree with a distinct stem.

Brachiate - a. Spreading with branches suggesting arms.

Branch - n. A lateral division of the stem or axis of growth.

Branchlet - n. The ultimate division of a branch.

Bud - n. An embryonic axis with its appendages.

Bulb - n. A modified bud usually underground; **imbricated** - with scaly modified leaves, as in the lily; **tunicated** - with modified leaves forming concentric layers around the bud, as the onion.

Bulbiferous - a. Bearing bulbs.

Bulbil - n. A bulb arising from the mother bulb.

Bulbet - n. A little bulb produced in the leaf axil, inflorescences, or other unusual places.

Bulbose - a. Having bulbs or the structure of a bulb.

Caespitose - a. Growing in tufts.

Caudex - n. The woody base of a perennial plant.

Caulescent - a. More or less stemmed or steam-bearing; having an evident stem above ground.

Cauliculous - a. With a small stem.

Cauline - a. Pertaining or belonging to the stem.

Cladophyll - n. A branch assuming the form and function of a leaf; a cladode.

Corm - n. A solid bulblike structure, usually subterranean, as the "bulb" of *Gladiolus*.

Cormel - n. A corm arising from a mother corm.

Crown - n. Corona; the base of a tufted, herbaceous, perennial grass; the hard ring or zone at the summit of the lemma of some species of *Stipa*; the part of a stem at the surface of the ground; a part of a rhizome with a large bud, used in propagation.

D. B. H. - n. Diameter breast high.

Deliquescent-a. Dissolving or melting away; said of a stem that loses itself by repeating branching; opposed to excurrent.

Digonous - a. Two-angled, as the stems of some *Cacti*.

Excurrent - a. With the stem remaining central and other parts being regularly disposed around it; running through to the apex.

Gemma - n. A bud or a body analogous to a bud capable of producing a new plant.

Haplocaulous - a. Having a simple unbranched stem.

Infra-axillary - a. Below the axil, sub-axillary.

Internode - n. The part of a stem between two successive nodes.

Melanoxylon - n. Black wood.

Nodal - a. Pertaining to the node.

Node - n. That point on a stem which normally bears a leaf or leaves.

Nodiferous - a. Bearing nodes.

Nodose - a. With nodes.

Ramal - a. Pertaining to a branch.

Rameal - a. See Ramal.

Rhizome - n. A dorsiventral, rootlike, underground stem which produces roots and shoots; rootstock.

Sapwood - n. The new wood in an exogenous tree, so long as it is pervious to water; the alburnum.

Sautellus - n. A bulblet, such as those of *Lilium tigrinum*.

Scapose - a. Bearing or resembling a scape.

Semester ring - n. The ring produced in the wood of many tropical trees in consequence of two periods of growth and rest in a year.

Stem - n. The main axis of a plant, leaf-bearing as distinguished from the rootbearing axis.

Stems, subterranean - pl. n. Rhizomes, tubers, bulbs, and corms.

Stipitate - a. With a stipe.

Sub-axillary - a. Borne below the axil.

Supra-axillary-a. Borne above the axil, super-axillary.

Triquetrous - a. With three salient angles and three concave faces.

Tuber-n. A short thickened branch of a subterraean stem, beset with buds or "eyes."

Tubercle - n. A little tuber.

Surfaces and Vestures

Shiny surfaces

- **Glittering** - a. With luster as from a polished surface which is not uniform.
- **Illustrous** - a. Bright, brilliant, lustrous.
- **Lucid** - a. Shiny, bright.
- **Lustrous** - a. Glossy, shiny.
- **Micaceous** - a. Glittering, sparkling, mica-like.
- **Nacreous** - a. With pearly luster.
- **Nitid** - a. Smooth and clear, lustrous, glittering.
- **Vernicose** - a. Shiny, as though varnished.

Smooth surfaces (without hairs, spines, bristles, or scales)

Regular

- **Alepidote** - a. Destitute of scurf or scales.
- **Bloom** - n. The white, waxy, or pruinose covering on many fruits, leaves, and stems.

Glabrate - a. Nearly glabrous, or becoming glabrous with maturity or age.
Glabrous - a. Smooth, devoid of pubescence or hair of whatsoever form.
Glaucous - a. Covered with a "bloom" or a whitish substance that rubs off, as of a plum or cabbage leaf.
Laevigate - a. Smooth, as if polished.
Naked - a. Wanting its usual covering, as without pubescence.
Pruinose - a. Having a waxy powdery secretion on the surface, a "bloom."
Unctuous - a. Having a surface which feels greasy.
Vernicose - a. Shiny, as though varnished.
Viscid - a. Sticky from a tenacious coating or secretion.

Irregular

Alveolate - a. Honey-combed.
Areola - n. A small area marked out on a surface.
Areolate - a. With areola.
Bullate - a. With surface blistered or puckered, as the leaf of a Savoy cabbage.
Canaliculate - a. Channeled longitudinally.
Colliculose - a. Covered with little round elevations or hillocks.
Corrugate - a. Wrinkled.
Faveolate - a. Honey-combed, alveolate.
Fluted - a. Regularly marked by alternating ridges and groovelike depressions.
Foveolate-a. Marked with small pitting.
Furrowed-a. With longitudinal channels or grooves, sulcate, striate on a large scale.
Lacuno-rugose - a. Marked with irregular wrinkles, as the stone of a peach.
Mammiform - a. Breast-shaped, conical with rounded apex.
Mammilla - n. A nipple or teat.
Mammillate - a. Having teat-shaped processes.
Mammose - a. Having teat-shaped processes.
Nodulose - a. With little knobs or knots.
Papilla (pl. **papillae**) - n. A minute nipple-shaped projection.
Papillary - a. Resembling papillae.
Papillose - a. Bearing papillae.
Pitted - a. Marked with small depressions, punctate.
Pustulose - a. Blistery, furnished with pustules or irregular raised pimples (not as roughened as papillose).
Rugose - a. Wrinkled, as leaf surface with sunken veins.
Rugulose - a. Somewhat wrinkled.
Scrobiculate - a. Marked by minute or shallow depressions, pitted.
Striate - a. With fine grooves, ridges, or lines of color.
Sulcate - a. Grooved or furrowed lengthwise.
Tesselate - a. Having the surface marked with square or oblong depressions.
Thelephorous - a. Covered with nipple-like prominences.
Verrucose - a. Covered with wartlike elevations.
Wrinkled - a. Rugose, irregularly creased.

Granular and scaly surfaces

Farinose - a. Covered with a mealiness.
Fornix (pl. **fornices**) - n. A small arched scale.

Granular – a. Covered with very small grains; minutely or finely mealy.
Grumose – a. Crumby.
Lepis (pl. **lepides**) – n. A scale, usually attached by its center.
Lepidote – a. Beset with small scurfy scales.
Pulveraceous – a. Covered with a layer of powdery granules.
Ramentum (pl. **ramenta**)–n. Used in the plural for the thin chaffy scales of the epidermis, as the scales of many ferns.
Scobinate – a. Rough as though rasped.
Scurf – n. Small branlike scales on the epidermis.
Scurfy – a. Covered with small scales.
Squamaceous – a. Scaly.
Squamate – a. Furnished with scales.
Squamose – a. Squamate, full of scales.
Stellate scales – pl. n. Trichomes, discs borne by their edge or center.
Tartareous – a. With a loose or rough crumbling surface, as some lichens.

Hairy surfaces

Straight hairs

Canescent – a. Gray-pubescent or becoming so.
Cilium (pl. **cilia**) – n. Used generally in the plural to desginate marginal hairs.
Ciliate – a. Said of a margin fringed with hairs.
Comose – a. Bearing a tuft or tufts of hair.
Crinus – n. A stiff hair on any part.
Down – n. Soft pubescence; the pappus of such plants as thistles.
Glochidiate – a. Pubescent with barbed bristles.
Hair – n. An outgrowth of the epidermis consisting of one to several cells.
Hirsute – a. With stiff or bristly hairs.
Hirsutulous – a. Slightly hirsute.
Hirtellous – a. Softly or minutely hirsute or hairy.
Hispid – a. Beset with rough hairs or bristles.
Hispidulous – a. Somewhat or minutely hispid.
Hoary – a. Covered with a close white or whitish pubescence.
Multiciliate – a. With many cilia.
Piliferous – a. Bearing hairs, or tipped with hairs; hair-pointed.
Pilose – a. With soft hairs.
Plumose – a. Pubescent in a manner simulating a feather or a plume.
Puberulent – a. Somewhat or minutely pubescent.
Pubescence – n. The hairiness of plants.
Pubescent – a. Covered with short soft hairs; down.
Scabrous – a. With short bristly hairs; rough to the touch.
Sericeous – a. Silky, clothed with closely appressed, soft, straight pubescence.
Silky – a. Said of a condition produced by a covering of soft, appressed, fine hairs; sericeous.
Strigose – a. Beset with sharp-pointed, appressed, straight, and stiff hairs or bristles; hispid.
Trichome – n. Any hairlike outgrowth of the epidermis, as a hair or bristle.
Tufted – a. Cespitose, comose, having a small cluster of hairs.

Velutinous – a. Velvety, due to a coating of fine, soft hairs.
Velumen – n. Close, short, soft hair.
Villose – a. Covered with long, silky, straight hairs.

Interwoven hairs

Arachnoid – a. Cobwebby, composed of slender entangled hairs; spider-like.
Eriophorous – a. Wool-bearing, densely cottony.
Felted – a. Matted with intertwined hairs.
Floccus (pl. **flocci**) – n. A lock of soft wool or hair.
Floccose – a. Bearing flocci.
Flocculose – a. Like wool.
Gossypine – a. Cottony, flocculent; like the hairs on the seeds of *Gossypium.*
Holosericeous – a. Woolly-silky.
Indument – n. Hairy or pubescent with rather heavy covering.
Lanate – a. Clothed with woolly and interwoven hairs.
Lanose – a. Woolly.
Lanuginose – a. Woolly or cottony.
Manicate – a. Said of pubescence so dense and interwoven that it may be stripped off.
Panniform – a. Having the appearance or texture of felt or woolen cloth.
Pannose – a. Having the appearance or texture of felt or woolen of very close texture.
Tomentose – a. With tomentum; densely woolly or pubescent; with matted soft woollike hairiness.
Tomentulose – a. Somewhat or delicately tomentose.
Tomentum – n. A densely matted pubescence.
Velutinous – a. Velvety, due to a coating of fine soft hairs.
Web – n. A cluster of slender, soft hairs.
Woolly – a. Lanate, tomentose, clothed with long and tortuous or matted hairs.

Hooked hairs

Aduncate – a. Bent or crooked, as a hook.
Aduncous – a. Hooked.
Uncinate – a. Hooked, bent at the tip like a hook.

Branched hairs

Candelabra hairs – pl. n. Stellate hairs in two or more tiers.
Stellipilous – a. With stellate hairs.
Stellate – a. Starry, often said of hairs that have radiating branches from base or of separate hairs similarly aggregated.

Spiny or prickly surfaces

Acantha – n. Thorn.
Aculeate – a. With prickles.
Aculeolate – a. Beset with small prickles.
Asperate – a. Rough.
Asperous – a. Rough or harsh to the touch.
Barbed – a. Furnished with retrorse projections.
Barbellate – a. Finely barbed.
Barbulate – a. Finely bearded.
Bearded – a. Bearing or furnished with long or stiff hairs.
Bristle – a. A stiff hair.
Bristly – a. Bearing stiff strong hairs.
Echinate – a. Armed with prickles.
Exasperate – a. Rough, with hard, projecting points.

Glochid - n. A barbed hair or bristle.

Glochideous - a. Pubescent with barbed bristles.

Muricate - a. Rough with short, hard points.

Muriculate - a. Very finely muricate.

Pungent - a. Ending in a rigid and sharp point.

Scabridulous - a. Slightly rough.

Scabrous - a. With short bristly hairs; rough to the touch.

Setigerous - a. Bristly, bristle-bearing.

Setose - a. Bristly, beset with bristles.

Setulose - a. With minute bristles.

Spiculose - a. With a surface covered with fine points.

Spinescent - a. Ending in a spine or sharp point.

Spinose - a. Spinelike; with spines or thorns.

Spinulose - a. With small spines or spinules.

Spiny - a. Beset with spines.

Miscellaneous surfaces

Cilium (pl. **cilia**) - n. Used generally in the plural to designate marginal hairs.

Ciliate - a. Said of a margin fringed with hairs.

Ciliolate - a. Minutely ciliate.

Derma (pl. **dermata**) - n. Surface of an organ, bark, rind, or skin.

Fimbria - n. A fringe.

Fimbriate - a. Fringed, the hairs longer or coarser as compared with ciliate.

Fimbrillate - a. With a minute fringe.

Malpighiaceous hairs - n. Hairs which are straight and appressed but attached, by the middle, frequent in the family *Malpighiaceae*.

Pelta - a. A bract attached by its middle as in peppers.

Piloglandulose - a. Bearing glandular hairs.

Process - n. Any projecting appendage.

Punctate - a. Dotted with depressions or punctures.

Stinging hair - n. A hollow hair seated on a gland which secretes an acid substance, as in nettles.

Tentacle - n. A sensitive glandular hair, as those on the leaf of *Drosera*.

Trap-hairs - n. Special hairs which confine insects in certain flowers until pollination is effected.

Tricholoma - n. With the edge or border furnished with hairs.

Texture

Callose - a. Hard and thick in texture.

Cartilagineous - a. Like cartilage or gristle.

Cereous - a. Waxy.

Chaffy - a. With small membranous scales.

Chartaceous - a. Having the texture of writing paper.

Coriaceous - a. Leather-like.

Corneous - a. Horny, with a horny texture.

Crustaceous - a. Of hard and brittle texture.

Crystalline - a. Resembling a crystal.

Fibrous - a. Having much woody fiber, as the rind of a coconut.

Flaccid - a. Withered and limp, flabby.

Fleshy - a. Succulent.

Fragile – a. Weak, easily broken.
Frutescent – a. Becoming shrubby.
Fruticose – a. Shrubby or shrub-like in the sense of being woody.
Gelatinous – a. Jelly-like.
Glutinous – a. Sticky.
Granular – a. Composed of grains; divided into little knots or tubercles.
Herbaceous – a. With the texture, colors, and properties of a herb, not woody.
Hyaline – a. Thin and translucent or transparent.
Indurated – a. Hardened.
Ligneous – a. Woody.
Membranaceous – a. The same as membranous.
Membranous – a. Thin, more or less flexible, and translucent.
Mucilaginous – a. Slimy, composed of mucilage.
Oleaginous – a. Oily and succulent.
Oleiferous – a. Oil-bearing.
Oleraceous – a. Herbaceous.
Osseous – a. Bony.
Ossified – a. Made hard as bone, as the stones of drupes, such as the peach or plum.
Paleaceous – a. Chaffy.
Papery – a. Having the texture of paper, chartaceous.
Papyraceous – a. Papery, white as paper.
Resinous – a. Like, or pertaining to resin.
Sap – n. The juice of a plant; the fluid contents of cells and young vessels consisting of water and salts absorbed by the roots and distributed through the plant.
Saponaceous – a. Soapy, slippery to the touch.
Scarious – a. Thin, dry, and membranous, often more or less translucent.
Scleroid – a. Having a hard texture.
Sebaceous – a. Like lumps of tallow.
Spumose – a. Frothy.
Suberous – a. Corky in texture.
Succulent – a. Juicy or pulpy.
Suffruticose – a. Shrubby at the base.
Viscid – a. Sticky from a tenacious coating or secretion.

Time and Season

Aestival – a. See Estival.
Allochronic species – pl. n. Species which do not belong to the same time level, as opposed to contemporary or synchronic.
Allogenous flora – pl. n. Relic plants of an earlier prevailing flora and environment; epibiotic plants.
Annotinous – a. A year old, or in yearly growths.
Annual – a. Of one year's duration; completing its life cycle in one year.
Asyngamic – a. Unable to cross by reason of differences in time of flowering.
Autumnal – a. Of or pertaining to autumn, flowering in autumn.
Biennial – a. Of two seasons' duration from seed to maturity and death.
Crepuscular – a. Of or pertaining to twilight.
Diplobiont – n. A plant flowering or fruiting twice each season.
Diurnal – a. Occurring in the daytime, sometimes used for ephemeral.
Efflorescence – n. The season of flowering; anthesis.
Ephemer – n. A flower which closes after a short term of expansion.
Ephemeral – a. Persisting for one day only, as flowers of spiderwort.
Estival – a. Of or pertaining to summer.
Frutescence – n. The time of maturity of fruit.
Hemeranthous – a. Day-flowering.

Hibernaculum – n. The winter resting part of a plant, as a bud or underground stem.
Hibernal – a. Hibernating, relating to winter.
Hibernation – n. Passing the winter in a dormant state.
Hiemal – a. Relating to winter.
Horary – a. Lasting an hour or two, as the petals of *Cistus*.
Hyemal – a. See Hiemal.
Matutinal – a. Pertaining to morning; plants flowering early, as *Ipomoea purpurea*.
Menstrual – a. Lasting for a month or so.
Nocturnal – a. Occurring at night, or lasting one night only.
Nox (pl. **noctes**) – n. Night.
Nyctanthous – a. Said of night-flowering plants.
Nyctigamous – a. With flowers which open at night and close by day; pollinated at night.
Nyctitropic – a. Turning in response to darkness.
Nyctitropism – n. The act of assuming the sleep position.
Paraheliotropism – n. Diurnal sleep, the movement of leaves to avoid the effect of intense sunlight.
Perennation – n. A lasting or perennial state.
Perennial – a. Said of a plant which lasts several years, not perishing normally after one flowering and fruiting.
Photeolic – a. Pertaining to the "sleep" of plants.
Precocious – a. Appearing or developing very early, as aments in *Salix* expanding before the leaves.
Prevernal – a. Of early spring.
Serotinal – a. Produced late in the season, or the year, as in autumn.
Serotinous – a. Produced or occurring late in the season.
Sleep – n. The response of plants, with changes in position of organs such as leaves due to the absence of light.
Sleep movement – n. Positions taken by leaves during the night, nyctitropic movement.
Synchronic species – pl. n. Species which belong to the same time level, contemporary.
Therophyllous – a. Producing leaves in summer, deciduous-leaved.
Therophyte – n. A plant which completes its development in one season, its seeds remaining latent during the hot season; an annual.
Trimestris – a. Of three months, as lasting that time or maturing in it.
Vernal – a. Pertaining to spring.
Vespertine – a. Appearing or expanding in the evening.

Tropisms

Tropism is the innate tendency of an organism to react in a definite manner to an external stimulus.

Anemotropism – n. The tropic response of organisms to wind and air currents.
Aphaptotropism – n. The state of not being influenced by touching stems or other surfaces.
Apheliotropism – n. The act of turning away from the sun; negative photropism.
Aphercotropism – n. The act of turning away from an obstruction.
Aphototropism – n. The act of turning away from the light.
Autonyctitropism – n. Regularly assuming the position usual during the night.
Autoörthotropism – n. The tendency of an organ to grow in a straight line forward.

Barotropism – n. The response of an organism to changes in barometric pressure.

Chemotropism – n. Curvature in response to chemical stimuli.

Cryotropism – n. Movements induced by cold or frost.

Diaphototropism – n. The act of turning at right angles to incident light, as the leaves of some plants.

Diatropism – n. The act of organs placing themselves crosswise to an operating stimulus.

Dromotropism – n. The tropic movement of climbing plants which results in their spiral growth.

Edaphotropism – n. Tropic responses to the soil.

Epigeotropism – n. Tropism resulting in growth on the surface of the soil.

Geodiatropism – n. A function which places an organ at right angles to the force of gravity.

Heliotropism – n. The act of turning in response to the sun.

Homalotropism – n. The act of turning to a horizontal position.

Hydrotropism – n. The act of turning in response to the influence of water.

Nyctitropism – n. The tendency of certain plant organs, as leaves, to assume special "sleeping" positions or to make curvatures under the influence of darkness.

Ombrotropism – n. Tropic responses of organisms to the stimulus of rain.

Orthoösmotropism – n. The act of assuming an erect position due to osmotic action.

Orthotropism – n. Growth in a vertical position.

Paraheliotropism – n. Diurnal sleep; movements of leaves to avoid the effects of intense sunlight.

Paraphototropism – n. The act of turning at right angles to the incident light.

Phototropism – n. The act of turning in response to light.

Polytropism – n. The act of leaves placing themselves vertically and meridionally, the two surfaces facing east and west respectively.

Prohydrotropism – n. The act of turning toward a source of moisture.

Rheotropism – n. The act of turning in response to a current of water.

Selenotropism – n. Movements of plants caused by the light of the moon.

Siotropism – n. Response to shaking, as with *Mimosa.*

Stenotropism – n. A condition with narrow limits of adaptations to varied conditions.

Stereotropism – n. Response to contact stimuli.

Telotropism – n. The act of turning to one stimulus to the exclusion of all others.

Thermotropism – n. The act of turning in response to heat.

Thigmotropism – n. The act of turning in response to a mechanical stimulus.

Topotropism – n. The act of turning toward the place from which a stimulus comes.

Traumatropism – n. The sensitiveness of certain plant organs to wounds.

Zenotropism – n. Negative geotropism.

Type Terminology

Agrotype – n. An agricultural race.

Biotype – n. A group of individuals all of one genotype.

Chirotype – n. The specimen on which a manuscript name is based.

Clastotype – n. A fragment from the original type.

Cotype – n. An additional or associate type specimen from which a taxon is described.

Generitype – n. The type species of a genus.

Genotype – n. The type of a genus, the species upon which the genus was established.

Holotype – n. The one specimen or other element used by the author of the name, or designated by him, as the nomenclatural type (i.e., the element to which the name of the taxon is permanently attached).

Icotype – n. A type serving for identification, but not previously used in literature.

Isotype – n. A specimen believed to be a duplicate of the holotype.

Lectotype – n. A specimen or other element selected from the original material to serve as the nomenclatural type, when the holotype was not designated at the time of publication, or when the holotype is missing.

Logotype – n. A type determined historically from two or more original species.

Merotype – n. A specimen collected from the original type in cultivation by means of vegetative reproduction.

Metatype – n. A specimen from the original locality, recognized as authentic by the describer himself.

Mimotype – n. Forms distinctly resembling each other, fulfilling similar functions, and thus representing each other in different floras.

Monotypic – a. Having only one exponent, as a genus with but one species.

Necrotype – n. A form that formerly existed but is now extinct.

Neotype – n. A specimen selected to serve as the nomenclatural type of a taxon in a situation when all material on which the taxon was based is missing.

Paratype – n. A specimen cited with the original description other than the holotype.

Phenotype – n. A group of individuals of similar appearance but not necessarily of similar genetic constitution.

Phototype – n. A photograph of a type specimen.

Proterotypes – pl. n. Primary types; all specimens which have served as a basis for descriptions and figures of organisms; further divided into holotype, cotype (or syntype), paratype, lectotype, and chirotype.

Prototype – n. The assumed ancestral form from which the descendents have become modified.

Spermotype – n. A specimen cut from a seedling grown from the original type.

Syntype – n. One of two or more specimens or elements used by an author when no holotype is designated, or one of two or more specimens simultaneously designated as type.

Tectoparatype – n. A specimen selected to show the microscopic structure of the original type of a species or genus.

Topotype – n. A specimen of a named species from the type locality.

Type specimen – n. The original specimen from which a description is written.

Typical – a. In classification, conforming to the originally described specimen.

Unitypic – a. See Monotypic.

Vernation

Vernation deals with the disposition of foliage leaves in the bud. It does not treat of the insertion of the leaves on the axis as this comes under phyllotaxy. The disposition of floral parts in the bud is treated under Aestivation.

Circinate – a. Coiled from the top downward, as the leaves of *Frosera* and the fronds of true ferns.

Complicate – a. Folded upon itself.

Conduplicate – a. Folded lengthwise, or doubled up flat on the midrib, the upper face of the leaf always within, as in *Magnolia*.

Convolute – a. Rolled up from one margin, one margin on the inside and the other on the outside.

Equitant – a. Folded over as if astride; used for conduplicate which enfold each other in two ranks, as in *Iris*.

Involute – a. With both margins rolled toward the midrib on the upper surface, as the leaves of water lily and violets.

Plicate – a. Folded on the several ribs in the manner of a closed fan. It occurs in palmately veined leaves, as in maple and currant.

Reclinate (inflexed) – a. With the upper part bent on the lower, or the blade on the petiole, as in *Liriodendron*.

Revolute – a. With both margins rolled toward the midrib on the lower face, as the leaves of *Azalea*.

Some Specific Epithets With Their Meanings

The specific epithet is the second element in a scientific name. It may be a noun (in the nominative or the genitive), or an adjective. When adjectival in form, and not used as a substantive, it agrees in gender with the generic name. The epithets are listed in the masculine gender, followed by the feminine and neuter endings. Present participles, which have the same endings, are indicated below.

A

abbreviatus, -a, -um — abbreviated, shortened.
abietinus, -a, -um — like the abies; or firlike.
abruptus, -a, -um — abrupt.
absinthioides, -es, -es absinthalike.
abyssinicus, -a, -um — Abyssinian.
acantha (noun) — thorn.
acanthocomus, -a, -um — spiny-haired or crowned.
acaulis, -is, -e — stemless.
accumulatus, -a, -um — accumulated.
acephalus, -a, -um — headless.
acerbus, -a, -um — harsh or sour.
acerifolius, -a, -um — maple-leaved.
aceroides, -es, -es — maple-like.
acetosus, -a, -um — acetic, sour.
acicularis, -is, -e — needle-like.
acidissimus, -a, -um — very sour.
acidus, -a, -um — acid, sour.
acinaceus, -a, -um — scimitar- or saber-shaped.
acris, -is, -e — acid, sharp.
acrostichoides, -es, -es — acrostichum-like.
acrotrichis, -is, -e — hairy-tipped.
aculeatus, -a, -um — with prickles.
acuminatus, -a, -um — tapering at the end.
acutifolius, -a, -um — with sharp leaves.
acutus, -a, -um — acute, sharp-pointed.
adenophorus, -a, -um — gland-bearing.
adenophyllus, -a, -um — glandular-leaved.
adiantifolius, -a, -um — adiantum-leaved.
admirabilis, -is, -e — admirable, noteworthy.

adnatus, -a, -um — adnate, joined to.
adpressus, -a, -um — pressed against.
adscendens (pres. part.) — ascending.
adsurgens (pres. part.) — ascending.
aduncus, -a, -um — hooked.
adventivus, -a, -um — adventitious.
advenus, -a, -um — newly arrived, adventive.
aerius, -a, -um — aërial.
aestivalis, -is, -e — summer.
aestivus, -a, -um — summer.
affinis, -is, -e — related (to another species).
agavoides, -es, -es — agave-like.
agglomeratus, -a, -um — heaped up, crowded together.
aggregatus, -a, -um — aggregate, clustered.
agrestis, -is, -e — growing wild.
agrostoides, -es, -es — similar to *Agrostis.*
airoides, -es, -es — like *Aira.*
alaris, -is, -e — wing-shaped.
alatus, -a, -um — winged.
albescens (pres. part.) — whitish, becoming white.
albicans (pres. part.) — whitening.
albicaulis, -is, -e — white-stemmed.
albidus, -a, -um — white.
albiflorus, -a, -um — with white flowers.
albipilosus, -a, -um — with white shaggy hair.
albispinus, -a, -um — white-spined.
albus, -a, -um — white.
alcicornis, -is, -e — elk-horned.
alexandrinus, -a, -um — of or pertaining to Alexandria (Egypt).
alliaceus, -a, -um — of the alliums, garlic-like, usually connoting odor.
almus, -a, -um — bountiful.
alnifolius, -a, -um — alder-leaved.
alopecuroides, -es, -es — foxtail-like.
alpestris, -is, -e — nearly alpine.
alpicolus, -a, -um — living in high mountains.
alpinus, -a, -um — alpine.
alternifolius, -a, -um — with alternate leaves.
altifolius, -a, -um — tall-leaved.
altissimus, -a, -um — very tall.
altus, -a, -um — high.
alveatus, -a, -um — honey-combed.
alveolaris, -is, -e — honey-combed.
amabilis, -is, -e — lovely.
amarus, -a, -um — bitter.
amazonicus, -a, -um — of the Amazon River region.
ambigens (pres. part.) — ambiguous.
ambiguus — doubtful.
amblyodus, -a, -um — blunt-toothed.
amentum (noun) — catkin.
amethystinus, -a, -um — amethystine, violet-colored.
amoenus, -a, -um — pleasant, charming.
amphibius, -a, -um — amphibious, growing on land or in water.
amplexicaulis, -is, -e — stem-clasping.
amplexifolius, -a, -um — leaf-clasping.
amurensis, -is, -e — of the Amur River region.
amygdaloides, -es, -es — almond-like.
anacanthus, -a, -um — without spines.
anceps, -ceps, -ceps — two-headed or two-edged.
aneurus, -a, -um — nerveless.
anfractuosus, -a, -um — twisted.
angelicus, -a, -um — English.
angularis, -is, -e — angular.
angustifolius, -a, -um — with narrow leaves.
angustus, -a, -um — narrow.
annotinus, -a, -um — year-old.
annularis, -is, -e — ring-shaped.
annuus, -a, -um — annual, yearly.
anomalus, -a, -um — unusual, out of the ordinary.

anserinus, -a, -um —of a goose.
antarcticus, -a, -um — of the Antarctic regions.
apertus, -a, -um — uncovered, bare, open.
apetalus, -a, -um — without petals.
aphyllus, -a, -um — without leaves.
apiculatus, -a, -um — tipped with a point.
apodus, -a, -um — footless.
appendiculatus, a, -um — appendaged.
applatanatus, -a, -um — flattened.
apterus, -a, -um — wingless.
aquaticus, -a, -um — aquatic.
aquatilis, -is, -e — aquatic.
aquifolius, -a, -um — with pointed leaves.
aquila (noun) — eagle.
arabicus, -a, -um — Arabian.
arachnifer, -era, -erum — spider-webby, web-bearing.
arachnoides, -es, -es — spider-like, cobwebby.
arborescens (pres. part.) — like a tree.
arboreus, -a, -um — treelike, pertaining to a tree.
articus, -a, -um — arctic, northern.
arcuatus, -a, -um — bowlike.
arenarius, -a, -um — of sand or sandy places.
arenicolus, -a, -um — sand-inhabiting.
areolatus, -a, -um — areolate, pitted.
argentatus, -a, -um — silvery, silvered.
argenteus, -a, -um — silvery.
argillaceus, -a, -um — clay, growing in clay, or clay-colored.
argophyllus, -a, -um — with silvery leaves.
argutus, -a, -um — sharp-toothed.
argyphyllus, -a, -um — silver-leaved.
argyreus, a, -um — silvery.
argyrocomus, -a, -um — silver-haired.
argyroneurus, -a, -um — with silver-colored nerves or veins.
aridus, -a, -um — arid, dry.
arietinus, -a, -um — like a ram's head.
aristatus, -a, -um — with an awn or beard.
armatus, -a, -um — armed (as with thorns).
armillaris, -is, -e — like a bracelet.
aromaticus, -a, -um — aromatic.
articulatus, -a, -um — with joints.
arundinaceus, -a, -um — reedlike.
arvensis, -is, -e — field-growing.
ascendens (pres. part.) — ascending.
asper, -era, -erum — rough.
asperifolius, -a, -um — with rough leaves.
asplenioides, -es, -es — asplenium-like.
assimilis, -is, -e — similar or like.
asteroides, -es, -es — aster-like.
ater, -atra, -atrum — coal-black.
atlanticus, a, -um — Atlantic, growing in Atlantic regions.
atratus, -a, -aum — blackened.
atriplicifolius, -a, -um — atriplex-leaved.
atrocaeruleus, -a, -um — dark cerulean or blue.
atrococcineus, -a, -um — dark scarlet.
atropuniceus, -a, -um — dark reddish-purple.
atropurpureus, -a, -um — dark purple.
atrorubens (pres. part.) — becoming dark red.
atrosanguineus, -a, -um — dark blood-red.
atrovirens (pres. part.) — becoming dark green.
attenuatus, -a, -um — attenuated, produced to a point.
augustissimus, -a, -um — very notable.
augustus, -a, -um — august, notable, majestic.
aurantiacus, -a, -um — orange-red.
aurantifolius, -a, -um — golden-leaved.

auratus, -a, -um — gold-colored.
aureolus, -a, -um — golden.
aureus, -a, -um — golden-yellow.
auricomus, -a, -um — golden-haired.
auriculatus, -a, -um — eared.
australis, -is, -e — southern.
austriacus, -a, -um — Austrian.
austrinus, -a, -um — southern.
autumnalis, -is, -e — autumnal, of autumn.
axillaris, -is, -e — axillary, pertaining to the axils.
azureus, -a, -um — sky blue.

B

baccatus, -a, -um — berry-like.
baccifer, -era, -erum — berry-bearing.
balsamifer, -era, -erum — balsam-bearing.
barbatus, -a, -um — bearded, barbed.
barbinervis, -is, -e — with nerves bearded.
barbinodis, -is, -e — barbed or bearded at nodes.
barbulatus, -a, -um — small-bearded.
basilaris, -is, -e — pertaining to the base or bottom.
basirameus, -a, -um — with branches low; toward bottom.
bellidifolius, -a, -um — beautiful-leaved.
bellus, -a, -um — nice, handsome.
benedictus, -a, -um — blessed.
benghalensis, -is, -e — of Bengal (E. India).
berolensis, -is, -e — of Berlin.
betaceus, -a, -um — of the beet, beetlike.
betulifolius, -a, -um — birch-leaved.
betulinus, -a, -um — birchlike.
bicarinatus, -a, -um — twice-keeled.
bicolor, -or, -or — two-colored.
bicornis, -is, -e — two-horned.
bidens, -ens, -ens — with two teeth.
bidentatus, -a, -um — two-toothed.
biennis, -is, -e — biennial.
bifidus, -a, -um — two-forked.
biflorus, -a, -um — two-flowered.
bifolius, -a, -um — with two leaves.
bifurcatus, -a, -um — twice-forked.
bigibbus, -a, -um — with two swellings or projections.
bijugus, -a, -um — yoked, two together.
bilobus, -a, -um — with two lobes.
bipinnatifidus, -a, -um — twice pinnately cut.
biserratus, -a, -um — twice-toothed.
bistortus, -a, -um — twice-twisted.
bisulcatus, -a, -um — two-grooved.
blandus, -a, -um — bland, mild, pleasant.
blephariglottis (noun) — fringed tongue.
blepharophyllus, -a, -um — fringed-leaved.
boliviensis, -is, -e — of Bolivia.
borealis, -is, -e — northern.
botryoides -es, -es — cluster-like, like the grape.
brachiatus, -a, -um — branched at right angles.
brachyandrus, -a, -um — short-stamened.
brachycarpus, -a, -um — short-fruited.
brachyphyllus, -a, -um — short-leaved.
brachystachys, -ys, -ys — short-spiked.
bracteatus, -a, -um — bearing bracts.
brasiliensis, -is, -e — of Brazil.
brevicaulis -is, -e — short-stemmed.
brevipes, -pes, -pes — short-footed.
brevis, -is, -e — short.
breviscapus, -a, -um — short-scaped.
brevisetus, -a, -um — short-bristled.
brunescens (pres. part.) — becoming brown.
brunneus, -a, -um — deep brown.

bufonius, -a, -um — pertaining to the toad.
bulbifer, -era, -erum — with bulbs.
bulbosus, -a, -um — bulbous.
bullatus, -a, -um — blistered, puckered.

C

cachemivicus, -a, -um — of Cashmere (Asia).
caerulescens (pres. part.) — becoming bluish.
caeruleus, -a, -um — blue.
caesius, -a, -um — bluish-gray.
caespitosus, -a, -um — growing in tufts.
caffer, caffra (noun) — Kafir (S. Africa).
calathinus, -a, -um — basket-like.
calcaratus, -a, -um — spurred.
calceiformis, -is, -e — shoe-shaped.
calceolus (noun) — small shoe.
calceus (noun) — shoe.
callianthemus, -a, -um — with beautiful flowers.
callicarpus, -a, -um — with beautiful fruit.
callistachyus, -a, -um — with beautiful spikes.
callizonus, -a, -um — beautiful-zoned.
callosus, -a, -um — callous, hardened.
calocomus, -a, -um — beautiful-haired.
calvus, -a, -um — bald, hairless.
calycinus, -a, -um — calyx-like.
calycosus, -a, -um — calyx-like.
campanula, (noun) — small bell.
campanularius, -a, -um — bell-shaped.
campanulatus, -a, -um — bell-shaped.
campestris, -is, -e — growing in fields or plains.
camphoratus, -a, -um — pertaining to camphor.
camphylacanthus, -a, -um — with crooked spines.
campschaticus, -a, -um — of Kamchatka.
camptchaticus, -a, -um — of Kamchatka.
canaliculatus, -a, -um — channeled, grooved.
canariensis, -is, -e — of the Canary Islands.
cancellatus, -a, -um — latticed.
candicans (pres. part.) — white-hairy or white-woolly.
candidus, -a, -um — white, shining.
canescens (pres. part.) — becoming grayish, becoming gray.
caninus, -a, -um — pertaining to a dog.
cannabinus, -a, -um — like *Cannabis* or hemp.
cantabricus, -a, -um — Cantabrian, of Cantabris (N. Spain).
cantonensis, -is, -e — of Canton (in S. China).
canus, -a, -um — ash-colored, hoary.
capensis, -is, -e — of the Cape (of Good Hope).
capillaris, -is, -e — hairlike.
capillus (noun) — hair.
capitatus, -a, -um — with a head.
capitellus (noun) — little head.
capreolatus, -a, -um — winding, twining.
capsularis, -is, -e — having capsules.
cardinalis, -is, -e — of cardinal-red color.
cardiopetalus, -a, -um—with heart-shaped petals.
carinatus, -a, -um — keeled.
carminatus, -a, -um — carmine.
carmineus, -a, -um — carmine.
carmosinus, -a, -um — crimson.
carneus, -a, -um — flesh-colored.
carnosus, -a, -um — fleshy.
carolinanus, -a, -um — pertaining to the Carolinas.
carolinus, -a, -um — pertaining to the Carolinas.
carpathicus, -a, -um — of the Carpathian region.
cashmerianus, -a, -um — of Cashmere (Asia).
catharticus, -a, -um — cathartic.

caucasicus, -a, -um — belonging to the Caucasus.
caudatus, -a, -um — with a tail.
caulescens (pres. part.) — becoming a stem.
caulialatus, -a, -um — wing-stemmed.
caulinus, -a, -um — belonging to the stem.
cayennensis, -is, -e — of Cayenne (French Guiana).
cellularis, -is, -e — cellular.
centifolius, -a, -um — hundred-leaved, many-leaved.
cepa (noun) — onion.
cephalatus, -a, -um — bearing heads.
cephalonicus, -a, -um — of Cephalonia (one of Ionian Islands).
cephalotes, -es, -es — headlike.
cerasifer, -era, -arum — bearing like a cherry.
ceratocaluis, -is, -e — horn-stalked.
cerealis, -is, -e — pertaining to Ceres or agriculture.
cereus, -a, -um — wax-colored.
cerifolius, -a, -um — wax-leaved.
cernuus, -a, -um — drooping, nodding.
chalcedonicus, -a, -um — of Chalcedon (on the Bosphorus).
chalepensis, -is, -e — of Aleppo (S.W. Asia).
chamaedrifolius, -a, -um — with leaves like a dwarf oak.
chamaedrys (noun) — dwarf oak.
chartaceus, -a, -um, — chartaceous papery.
cheilanthus, -a, -us — lip-flowered.
chilensis, -is, -e — belonging to Chile (S. America).
chiloensis, -is, -e — of Chile (S. America).
chionanthus, -a, -um — snow-flowered.
chloranthus, -a, -um — green-flowered.
chrysanthus, -a, -um — golden-flowered.
chrysocarpus, -a, -um — golden-fruited.
chrysocomus, -a, -um — golden-haired.
chrysolepis, -is, -is — golden-scaled.
chrysomallus, -a, -um — with golden wool.
chrysophyllus, -a, -um — golden-leaved.
chrysostomus, -a, -um — golden-mouthed.
cicutarius, -a, -um — of or like *Cicuta.*
cilianensis, -is, -e — hairy.
ciliatifolius, -a, -um — hairy leaves.
ciliatus, -a, -um — ciliate.
cilicicus, -a, -um — of Cilicia (S. E. Asia Minor).
ciliosus, -a, -um — ciliate, fringed.
cinctus, -a, -um — girdled.
cinereus, -a, -um — ash gray.
cinnabarinus, -a, -um — cinnabar-red.
cinnamomeus, -a, -um — cinnamon brown.
circinalis, -is, -e — coiled.
circinatus, -a, -um — coiled.
circularis, -is, -e — round.
cirrhosus, -a, -um — tendriled.
citratus, -a, -um — citrus-like.
citreus, -a, -um — lemon-colored.
citrinus, -a, -um — lemon yellow.
clandestinus, -a, -um — concealed.
clausus, -a, -um — shut, closed.
clavatus, -a, um — club-shaped.
claviformus, -a, -um — club-shaped.
coarctatus, -a, -um — appressed, crowded together.
cocciger, -era, -erum — berry-bearing.
coccineus, -a, -um — scarlet.
cochlearis, -is, -e — spoonlike.
coelestinus, -a, -um — sky blue.
coelestis, -is, -e — sky blue.
coerulescens (pres. part.) — becoming blue.
coeruleus, -a, -um — blue.
colchieus, -a, -um — of Colchis (eastern Black Sea region).
collinus, -a, -um — pertaining to a hill.

colonus, -a, -um — cultivated.

columellaris, -is, -e — pertaining to a small pillar or pedicel.

columnarius, -a, -um — columnar.

comans, -ans, -ans — with hair, or hairlike.

comatus, -a, -um — with hair.

communis, -is, -e — growing in common, gregarious.

commutatus, -a, -um — changing.

comosus, -a, -um — long-haired.

compactus, -a, -um — compact, dense.

complanatus, -a, -um — flattened.

complexus, -a, -um — encircled, embraced.

compressus, -a, -um — compressed.

conchaefolius, -a, -um — shell-leaved.

concinnus, -a, -um — neat, well made, elegant.

concolor, -or, -or — of the same color.

confertus, -a, -um — crowded.

confinis, -is, -e — bordered, bound.

conglomeratus, -a, -um — conglomerate, crowded together.

conoidus, -a, -um — conelike.

conspersus, -a, -um — scattered.

constrictus, -a, -um — constricted.

contortus, -a, -um — contorted, twisted.

contractus, -a, -um — narrowed.

convallis, -is, -e — valley.

convolvulus, -a, -um — winding, twining.

copallinus, -a, -um — like copal or resin.

cordatus, -a, -um — heart-shaped.

cordiformis, -is, -e — shaped like a heart.

coriaceus, -a, -um — like leather.

corniculatus, -a, -um — with horns, horned.

corniger, -era, -erum — horn-bearing.

cornutus, -a, -um — horny.

corollatus, -a, -um — corollate, like a corolla.

coronarius, -a, -um — suitable for wreaths.

coronatus, -a, -um — crowned.

corrugatus, -a, -um — corrugated, wrinkled.

corymbosus, -a, -um — corymb-like.

costatus, -a, -um — costate, ribbed.

crassifolius, -a, -um — thick-leaved.

crassipes, -es, -es — thick-footed.

crassus, -a, -um — thick.

crebrus, -a, -um — close, frequent, repeated.

crenatiflorus, -a, -um — crenate-flowered.

crenatus, -a, -um — crenellated, scalloped.

crenulatus, -a, -um — crenulate, finely scalloped.

creticus, -a, -um — of Crete (island of Crete).

crinitus, -a, -um — provided with long hair.

crispifolius, -a, -um — with leaves crisped or curled.

crispus, -a, -um — crisp, curled.

cristagalli (noun) — cockscomb.

cristatus, -a, -um — crested.

crocatus, -a, -um — saffron yellow.

croceus, -a, -um — saffron-colored, yellow.

cruciatus, -a, -um — cruciate, crosslike.

cruciformis, -is, -e — cross-shaped.

cruentus, -a, -um — blood red, bloody.

crusgalli (noun) — cockspur.

crustatus, -a, -um — encrusted.

cryptandrus, -a, -um — with hidden stamens.

crystallinus, -a, -um — crystalline.

ctenoides, -es, -es — comblike.

cucullatus, -a, -um — hooded.

cultriformis, -is, -e — shaped like a broad knife blade.

cuneatus, -a, -um — wedge-shaped.

cuneifolius, -a, -um — with wedge-shaped leaves.

cuneiformis, -is, -e — wedge-shaped.

cupreatus, -a, -um — coppery.

cupreus, -a, -um — copper-colored.

curtipedicellatus, -a, -um — short-pediceled.
curtipendulus, -a, -um — with short pendula.
curvatus, -a, -um — curved.
cuspidatus, -a, -um — sharp-pointed.
cyananthus, -a, -um — blue-flowered.
cyaneus, -a, -um — cornflower blue.
cyanophyllus, -a, -um — blue-leaved.
cyanthus, -a, -um — blue-flowered.
cylindricus, -a, -um — cylindrical.
cymosus, -a, -um — bearing cymes.

D

dactylifer, -era, -erum — finger-bearing.
dactyloides, -es, -es — finger-shaped.
dactylon (noun) — finger.
dahuricus, -a, -um — of Dahuria or Dauri (Trans-Baikal, Siberia).
damascenus, -a, -um — of Damascus.
dasycarpus, -a, -um — thick-fruited.
dasyphyllus, -a, -um — thick-leaved.
dasystachys, -ys, -ys — thick-spiked.
dauricus, -a, -um — of Dahuria or Dauria (Trans-Baikal, Siberia).
davuricus, -a, -um — of Dahuria or Dauria (Trans-Baikal, Siberia).
dealbatus, -a, -um — whitened, white-washed.
debilis, -is, -e — weak, frail.
decapetalus, -a, -um — ten-petaled.
deciduus, -a, -um — deciduous.
decipiens (pres. part.) — deceptive.
decorus, -a, -um — elegant, comely.
decumbens (pres. part.) — reclining.
decurrens (pres. part) — decurrent, running down the stem.
deflexus, -a, -um — bent downward.
deliciosus, -a, -um — delicious.
deltoides, -es, -es — triangular.
demersus, -a, -um — under the water.
demissus, -a, -um — low, weak.
dendroideus, -a, -um — treelike.
densus, -a, -um — dense.
dentatus, -a, -um — toothed.
denticulatus, -a, -um — with teeth.
denudatus, -a, -um — nude, naked.
depauperatus, -a, -um — stunted, dwarfed.
depressus, -a, -um — depressed.
deustus, -a, -um — burned.
diacanthus, -a, -um — two-spined.
diandrus, -a, -um — two-stamened.
dichotomiflorus, -a, -um — flowering in twos.
dichotomus, -a, -um — forked in pairs.
dicoccus, -a, -um — with two berries.
diffissus, -a, -um — split.
diffusus, -a, -um — loosely branching.
digitatus, -a, -um — finger-shaped.
digitus (noun) — finger.
dilitatus, -a, -um — dilated, expanded.
dimidiatus, -a, -um — halved.
dioicus, -a, -um — dioecious.
dipsacus, -a, -um — of the teasel or *Dipsacus*.
dipterocarpus, -a, -um — with two-winged carpel or fruit.
discolor, -or, -or — of different colors.
dissectus, -a, -um — dissected, deeply cut.
dissitiflorus, -a, -um — remotely or loosely flowered.
distachyus, -a, -um — two-spiked.
distichus, -a, -um — two-ranked; leaves or flowers in rank on opposite sides of stem.
diurnus, -a, -um — of the day, as day-flowering.
divaricatus, -a, -um — spreading, widely divergent.
divergens (pres. part.) — wide-spreading.

diversiflorus, -a, -um — diversely flowered, flowers variable.
diversifolius, -a, -um — with variable leaves.
divisus, -a, -um — divided.
dodecus, -a, -um — twelve.
dolabratus, -a, -um — mattock-shaped.
domesticus, -a, -um — indigenous, at home.
donax, -ax, -ax — reed or cane.
dracocephalus, -a, -um — dragon-head.
dracunculoides, -es, -es — tarragon- or dragon-like.
drepanophyllus, -a, -um — with sickle-shaped leaves.
dubius, -a, -um — doubtful.
dulcis, -is, -e — sweet.
dumosus, -a, -um — bushy.
duracinus, -a, -um — hard-berried.
duriusculus, -a, -um — somewhat hard or rough.

E

ebracteatus, -a, -um — without bracts.
eburneus, -a, -um — ivory white.
echinatus, -a, -um — prickly, like a headgehog.
edulis, -is, -e — edible.
effusus, -a, -um — very loose-spreading.
elasticus, -a, -um — elastic.
elatior, -ior, -ius — taller.
elatus, -a, -um — tall, high.
elegans, -ans, -ans — elegant.
elegantissimus, -a, -um — most elegant.
elevatus, -a, -um — elevated.
ellipsoidalis, -is, -e — elliptic.
ellipticus, -a, -um — elliptical.
elongatus, -a, -um — lengthened, long.
emarcidus, -a, -um — faded.
emarginatus, -a, -um — with a shallow notch at apex.
emersus, -a, -um — emersed.
enodis, -is, -e — without nodes.
ensatus, -a, -um — sword-shaped.
ensiformis, -is, -e — sword-shaped.
ensifolius, -a, -um — with sword-shaped leaves.
ephemericus, -a, -um — ephemeral, lasting but a day.
epigaeus, -a, -um — above the soil.
equus (noun) — horse.
erectus, -a, -um — upright.
eriacanthus, -a, -um — woolly-spined.
eriantherus, -a, -um — woolly- anthered.
erianthus, -a, -um — woolly-flowered.
ericoides, -es, -es — like *Erica,* heath-like.
eriocarpus, -a, -um — woolly-fruited.
eriopodus, -a, -um — having hairs at base.
erosus, -a, -um — erose, as if gnawed.
erubescens (pres. part.) — blushing.
erythrocarpus, -a, -um — red-fruited.
erythropus, -pus, -pus — red-footed, or red-stalked.
erythrosepalus, -a, -um — red-sepaled.
erythrosorus, -a, -um — with red sori.
esculentus, -a, -um — edible.
europaeus, -a, -um — European.
evertus, -a, -um — expelled, turned out.
exaltatus, -a, -um — exalted, verv tall.
excelsior, -ior, -ius — taller.
excelsus, -a, -um — tall, high.
excisus, -a, -um — excised, cut away.
excolor, -or, -or — colorless.
exiguus, -a, -um — little, small, poor.
eximius, -a, -um — distinguished, out of the ordinary, excellent.
exoticus, -a, -um — exotic, from another country.
expansus, -a, -um — expanded.
extensus, -a, -um — spread out or stretched out.

F

fabarius, -a, -um — pertaining to beans.
falcatus, -a, -um — sickle-shaped.
fallax, -ax, -ax — deceptive.
farinosus, -a, -um — farinaceous, mealy.
fasciatus, -a, -um — fasciate, coalesced.
fasciculatus, -a, -um — fascicled, clustered.
fastigiatus, -a, -um — fastigiate, branches erect and close together.
fastuosus, -a, -um — proud.
fatuus, -a, -um — foolish, simple, insipid.
fecundus, -a, -um — fertile.
femininus, -a, -um — feminine.
fenestralis, -is, -e — lattice-shaped.
fenestrellatus, -a, -um — with small window-like openings.
ferox, -ox, -ox — ferocious, very thorny.
ferrugineus, -a, -um — rusty, of the color of iron rust.
fertilis, -is, -e — fertile, fruitful.
fibrilla (noun) — small fiber.
fibrosus, -a, -um — fibrous, with fibers.
ficifolius, -a, -um — with leaves like a fig.
ficoides, -es, -es — figlike.
filamentosus, -a, -um — filamentous.
filicifolius, -a, -um — fern-leaved.
filifer, -era, -erum — with filaments or threads.
filifolius, -a, -um — thread-leaved.
filiformis, -is, -e — long, very slender, thread like.
fimbriatus, -a, -um — fringed, fimbriate.
firmus, -a, -um — strong, firm.
fissilis, -is, -e — cleft or split.
fistulosus, -a, -um — hollow-cylindrical.
flabellatus, -a, -um — with fan-like parts.
flabelliformis, -is, -e —fan-shaped.
flaccidus, -a, -um — flaccid, soft.
flagellaris, -is, -e — whiplike.
flagelliformis, -is, -e — whip-shaped.
flammeus, -a, -um — fire red, flame-colored.
flavens (pres. part.) — becoming yellowish.
flavescens (pres. part.) — becoming yellow.
flavicomus, -a, -um — yellow-wooled or yellow-haired.
flavispinus, -a, -um — yellow-spined.
flavissimus, -a, -um — deep yellow, very yellow.
flavus, -a, -um — yellow.
flexibilis, -is, -e — flexible.
flexicaulis, -is, -e — pliant-stemmed.
flexilis, -is, -e — pliant.
flexuosus, -a, -um — alternately bent in different directions.
floralis, -is, -e — floral.
florentinus, -a, -um — of Florence.
florepleno — with full or double flowers.
floribundus, -a, -um — free-flowering, blooming profusely.
floridanus, -a, -um — flowering, or from Florida.
floridus, -a, -um — flowering, full of flowers.
fluminensis, -is, -e — of a river.
fluitans (pres. part.) — floating.
fluvialis, -is, -e — fluvial.
foeminus, -a, -um — feminine.
foetidissimus, -a, -um — very fetid.
foetidus, -a, -um — having an offensive odor.
foliatus, -a, -um — with leaves.
foliosus, -a, -um — leafy.
follicularis, -is, -e — bearing follicles.
fontinalis, -is, -e — pertaining to a spring of water.
formosissimus, -a, -um — most or very beautiful.
formosus, -a, -um — beautiful, handsome.
foveolatus, -a, -um — pitted, with small depressions.

fragilis, -is, -e — fragile, brittle.
fragrans, -ans, -ans — fragrant, odorous.
fragrantissimus, -a, -um — very fragrant.
frigidus, -a, -um — cold, of cold regions.
frondosus, -a, -um — leafy.
fructigenus, -a, -um — fruitful.
frumentaceus, -a, -um — pertaining to grain (or corn).
frutescens (pres. part.) — becoming shrubby.
fruticosus, -a, -um — shrubby.
fucatus, -a, -um — painted, dyed.
fulgens, -ens, -ens — shining, glistening.
fulgidus, -a, -um — shining, glistening.
fuliginosus, -a, -um — sooty, black-colored.
fulvidus, -a, -um — slightly tawny.
fulvus, -a, -um — brownish-yellow.
funebris, -is, -e — funeral.
funiculatus, -a, -um — of a slender rope or cord.
furcatus, -a, -um — forked.
fuscus, -a, -um — brown, dusky.
fusiformis, -is, -e — spindle-shaped.

G

galeatus, -a, -um — helmeted.
galericulatus, -a, -um — helmet-like.
gallicus, -a, -um — of Gaul or France; also pertaining to a cock or rooster.
gandavensis, -is, -e — belonging to Ghent, Belgium.
geminatus, -a, -um — double or paired.
geminiflorus, -a, -um — twin-flowered.
geminispinus, -a, -um — twin-spined.
gemmatus, -a, -um — bearing buds.
gemmifer, -era, -erum — bud-bearing.
generalis, -is, -e — general, prevailing.
geniculatus, -a, -um — bent more or less like a knee.
germanicus, -a, -um — German.
gibberosus, -a, -um — humped, hunch-backed.
gibbiflorus, -a, -um — gibbous-flowered.
gibbosus, -a, -um — swollen on one side.
gibbus, -a, -um — gibbous, swollen on one side.
gibraltaricus, -a, -um — of Gibralter.
giganteus, -a, -um — gigantic, very large.
giganthes, -es, -es — giant-flowered.
glabellus, -a, -um — smoothish.
glaberrimus, -a, -um — very smooth, smoothest.
glabratus, -a, -um — somewhat glabrous.
glabrescens (pres. part.) — becoming smooth.
glabriflorus, -a, -um — smooth-flowered.
gladiatus, -a, -um — swordlike.
gladius (noun) — sword.
glandulifer, -era, -erum — gland-bearing.
glandulosus, -a, -um — glandular.
glaucescens (pres. part.) — becoming glaucous.
glaucophyllus, -a, -um — glaucous-leaved.
glaucus, -a, -um — glaucous, with a bloom.
globosus, -a, -um — like a globe.
globulus, -a, -um — like a small globe.
glomeratus, -a, -um — with dense or compact cluster or clusters.
gloriosus, -a, -um — glorious, superb.
glumaceus, -a, -um — with glumes or glumelike structures.
glutinosus, -a, -um — sticky.
gnaphalodes, -es, -es — like *Gnaphalium*.

gomphocephalus, -a, -um — club-headed.
goniatus, -a, -um — angled, cornered.
gossypinus, -a, -um — gossypium-like, cotton-like.
gracilis, -is, -e — slender.
gracillimus, -a, -um — very slender.
graecizans (pres. part.) — becoming widespread.
graecus, -a, -um — of Greece, Greek.
gramineus, -a, -um — grassy, grasslike.
graminifolius, -a, -um — with grasslike leaves.
grandidentatus, -a, -um — large-toothed.
grandiflorus, -a, -um — with large flowers.
grandifolius, -a, -um — with large leaves.
grandipunctatus, -a, -um — with large spots.
grandis, -is, -e — large, big.
granulatus, -a, -um — granular, covered with small grains.
granulosus, -a, -um — granulate.
gratiosus, -a, -um — agreeable.
gratissimus, -a, -um — very pleasing or agreeable.
graveolens (pres. part.) — strong scented.
griseus, -a, -um — gray.
grumosus, -a, -um — crummy.
gumminfer, -era, -erum — gum-bearing.
guttatus, -a, -um — with drops.
gymnocarpus, -a, -um — naked-fruited.

H

haemanthus, -a, -um — with blood-red flowers.
halepensis, -is, -e — of Aleppo (S. W. Asia).
halophilous, -a, -um — salt-loving.
hamatus, hamosus, -a, -um — hooked.
hastatus, -a, -um — spear-shaped.
hebecarpus, -a, -um — pubescent-fruited.
hederaceus, -a, -um — of or like the ivy (*Hedera*).
helianthoides, -es, -es — helianthus-like.
helodoxa (noun) — marsh beauty.
helvinus, -a, -um — yellowish.
helvolus, -a, -um — pale yellow.
herbaceus, -a, -um — herbaceous, not woody.
hesperius, -a, -um — of the west, or evening.
heteranthus, -a, -um — various-flowered.
heterocarpus, -a, -um — various-fruited.
heterolepis, -is, -is — with heterogeneous scales.
heterophyllus, -a, -um — with leaves of more than one shape.
hexagonus, -a, -um — six-angled.
hexapetalus, -a, -um — six-petaled.
hians (pres. part.) — open, gaping.
hibernicus, -a, -um — of Ireland.
hibernus, -a, -um — belonging to winter, wintry.
hiemalis, -is, -e — pertaining to winter.
himalicus, -a, -um — Himalayan.
hircinus, -a, -um — with a goat's odor.
hirsutissimus, -a, -um — very hairy.
hirsutus, -a, -um — bristly or prickly.
hirtellus, -a, -um — somewhat hairy.
hirtiflorus, -a, -um — hairy-flowered.
hirtipes, -es, -es — hairy-footed.
hirtus, -a, -um — hairy.
hispanicus, -a, -um — Spanish, of Spain.
hispidulus, -a, -um — somewhat bristly.
hispidus, -a, -um — with stiff hairs.
hollandicus, -a, -um — of Holland.
holosericeus, -a, -um — woolly-silky.

homolepis, -is, -is — with one kind of scales.
horizontalis, -is, -e — horizontal.
horridus, -a, -um — horrid.
hortensis, -is, -e — pertaining to a garden.
hortulanus, -a, -um — belonging to a garden, or gardens.
humifusus, -a, -um — sprawling on the ground.
humilis, -is, -e — dwarf, low growing.
hybridus, -a, -um — hybrid.
hyemalis, -is, -e — of winter.
hymenanthus, -a, -um — membranaceous-flowered.
hymenodes, -es, -es — membrane-like.
hymenosepalus, -a, -um — with membranous sepals.
hyperboreus, -a, -um — of the far North.
hypnoides, -es, -es — mosslike.
hypocrateriformis, -is, -e — salver-shaped.
hypogaeus, -a, -um — under the earth or soil.
hypoglaucus, -a, -um — glaucous beneath.
hypoleucus, -a, -um — Whitish, pale beneath.
hystrix, -ix, -ix — bristly, porcupine-like.

I

ibericus, -a, -um — of Iberia (the Spanish peninsula).
igneus, -a, -um — fire red.
ignis (noun) — fire.
ilicifolius, -a, -um — ilex-leaved, holly-leaved.
illustris, -is, -e — bright, brilliant, lustrous.
imberbis, -is, -e — without beards or spines.
imbricatus, -a, -um — overlapping like shingles.
immaculatus, -a, -um — immaculate, spotless.
immersus, -a, -um — under water.
impatiens, -ens, -ens — impatient.
imperator, -or, -or — commanding, imperious.
imperialis, -is, -e — imperial.
incanus, -a, -um — gray, hoary.
incarnatus, -a, -um — flesh-colored.
incertus, -a, -um — uncertain, doubtful.
incisifolius, -a, -um — cut-leaved.
incisus, -a, -um — incised.
incrassatus, -a, -um — thickened.
indentatus, -a, -um — indented.
indicus, -a, -um — Indian, of India or the East Indies.
indivisus, -a, -um — undivided.
induratus, -a, -um — hardened.
inequalis, -is, -e — unequal.
inermis, -is, -e — unarmed.
infectorius, -a, -um — used for dyeing, pertaining to dyes.
inferus, -a, -um — inferior.
inflatus, -a, -um — inflated.
infundibuliformis, -is, -e — funnel-shaped.
ingens, -ens, -ens — enormous.
inodorus, -a, -um — without odor, scentless.
inquinans, -ans, -ans — polluting, discoloring.
insectus, -a, -um — incised.
insignis, -is, -e — striking.
insititius, -a, -um — grafted.
insularis, -is, -e — pertaining to an island.
integerrimus, -a, -um — very entire.
integrifolius, -a, -um — with entire leaves.
intermedius, -a, -um — intermediate.
interruptis, -is, -e — not continuous, irregular.
introrsus, -a, -um — introrse, turned inward.
intumescens, -ens, -ens — tumid, swollen, enlarged, distended.
involucratus, -a, -um — with an involucre.
involutus, -a, -um — rolled inward.
ioensis, -is, -e — of Iowa (state in U.S.A.).

ionanthus, -a, -um — violet-flowered.
ischaemum (noun) — a blood styptic, an astringent.
isophyllus, -a, -um — equal-leaved.
italicus, -a, -um — Italian.
ixocarpus, -a, -um — sticky, or glutinous-fruited.

J

javanicus, -a, -um — Javan, of Java.
jubatus, -a, -um — crested, with a mane.
jucundus, -a, -um — delightful, pleasing.
jugosus, -a, -um — joined, yoked.
junceus, -a, -um — rushlike.
juniperinus, -a, -um — like juniper.

K

kamtschaticus, -a, -um — of Kamchatka.
kermesinus, -a, -um — crimson.
kewensis, -is, -e — belonging to Kew (Royal Botanic Gardens, Kew, England).

L

labiatus, -a, -um — with a lip.
labiosus, -a, -um — lipped.
labrosus, -a, -um — large-lipped.
laceratus, -a, -um — lacerate.
laciniatus, -a, -um — cut into narrow pointed lobes.
laciniosus, -a, -um — laciniose, laciniate, torn.
lactatus, -a, -um — milky.
lacteus, -a, -um — milky-white.
lactiflorus, -a, -um — with milk-colored flowers.
lacustris, -is, -e — growing in lakes.
ladanifer, -era, -erum — ladanum-bearing.
laetiflorus, -a, -um — with bright, or pleasing flowers.
laetivirens, -ens, -ens — bright green.
laetus, -a, -um — bright, vivid.
laevigatus, -a, -um — smooth.
laevipes, -pes, -pes — smooth-footed.
laevis, -is, -e — smooth.
lagemarius, -a, -um — of a bottle or flask.
lanatus, -a, -um — woolly.
lanceolatus, -a, -um — shaped like a lance.
lancifolius, -a, -um — lance-shaped leaves.
lanosus, -a, -um — woolly.
lanuginosus, -a, -um — woolly, downy.
lapideus, -a, -um — stony.
lasiandrus, -a, -um — pubescent-stamened.
lasianthus, -a, -um — woolly-flowered.
lasiocarpus, -a, -um — hairy- or woolly-fruited.
lasiolepis, -is, -is — woolly-scaled.
lasiopetalus, -a, -um — with petals rough-hairy.
lateralus, -a, -um — lateral, on the side.
lateritius, -a, -um — brick red.
latiflorus, -a, -um — broad-flowered.
latifolius, -a, -um — broad-leaved.
latiglumis, -is, -e — with wide glumes.
latimaculatus, -a, -um — broad-spotted.
latisquamus, -a, -um — broad-scaled.
latissimus, -a, -um — broadest, very broad.
laurifolius, -a, -um — laurel-leaved.
laurinus, -a, -um — laurel-like.
lavandulaceus, -a, -um — lavender-like.
laxiflorus, -a, -um — loose-flowered.
laxus, -a, -um — loose, open.
leianthus, -a, -um — smooth-flowered.
leiocarpus, -a, -um — smooth-fruited.
leiogynus, -a, -um — with a smooth pistil.

leiophyllus, -a, -um — smooth-leaved.
lenticularis, -is, -e — like a lens.
lentiginosus, -a, -um — freckled.
leopardinus, -a, -um — leopard-spotted.
lepidophyllus, -a, -um — with scaly leaves.
lepidotus, -a, -um — with small scurfy scales.
lepidus, -a, -um — graceful, elegant.
leprosus, -a, -um — scurfy.
leptanthus, -a, -um — thin-flowered.
leptocaulis, -is, -e — thin-stemmed.
leptocladus, -a, -um — thin-stemmed.
leptolepis, -is, -is — thin-scaled.
leptophyllus, -a, -um — thin-leaved.
leptopus, -us, -us — thin- or slender-stalked.
leptosepalus, -a, -um — thin-sepaled.
leucanthus, -a, -um — white-flowered.
leucocarpus, -a, -um — white-fruited.
leucocaulis, -is, -e — white-stemmed.
leucocephalus, -a, -um — white-headed.
leucodermis, -is, -e — white-skinned.
leuconeurus, -a, -um — white-nerved.
leucophyllus, -a, -um — white-leaved.
leucostachys, -ys, -ys — white-spiked.
leucotrichus, -a, -um — white-haired.
lignosus, -a, -um — like wood.
ligularis, -is, -e — ligulate, strap-shaped.
ligulatus, -a, -um — with a ligule.
limosus, -a, -um — of marshy or muddy places.
linearifolius, -a, -um — with long, slender leaves.
linearis, -is, -e — long and slender.
lineatus, -a, -um — lined, with lines or stripes.
linguiformis, -is, -e — tongue-shaped.
lingulatus, -a, -um — tongue-shaped.
linifolius, -a, -um — flax-leaved.
lithospermus, -a, -um — with stone-like seeds.
litoralis, -is, -e — growing on the seashore.
lividus, -a, -um — livid, bluish.
lobatus, -a, -um — lobed.
lobularis, -is, -e — lobed.
locularis, -is, -e — with locules.
longebracteatus, -a, -um — long-bracted.
longepedunculatus, -a, -um — long-peduncled.
longestylus, -a, -um — long-styled.
longiflorus, -a, -um — long-flowered.
longifolius, -a, -um — long-leaved.
longipes, -pes, -pes — long-footed.
longipetalus, -a, -um — long-petaled.
longiscapus, -a, -um — long-scaped.
longispicus, -a, -um — long-spiked.
longistipatus, -a, -um — long-stiped or long-stemmed.
longistylus, -a, -um — long-styled.
lophanthus, -a, -um — crest-flowered.
lorifolius, -a, -um — strap-shaped leaves.
loriformis, -is, -e — strap-shaped.
lucidus, -a, -um — shining, bright.
ludovicianus, -a, -um — of Louisiana.
lunatus, -a, -um — moon-shaped or crescent.
lupulinus, -a, -um — like hops.
luridus, -a, -um — dim yellow.
luteolus, -a, -um — yellowish.
lutescens (pres. part.) — yellowish or becoming yellow.
luteus, -a, -um — yellow.
lyratus, -a, -um — in the form of a lyre.

M

macilentus, -a, -um — lean, meager.
macradenus, -a, -um — long glands.
macranthus, -a, -um — with long or large flowers.
macrocarpus, -a, -um — with long or large fruits.
macrocephalus, -a, -um — large-headed.
macropetalus, -a, -um — large-petaled.
macrophyllus, -a, -um — large-leaved, long-leaved.
macropodus, -a, -um — large-footed.
macrorhizus, -a, -um — with large or long roots.
macrostylus, -a, -um — large-styled.
macrus, -a, -um — long or large.
maculatus, -a, -um — spotted, blotched, or stained.
magellanicus, -a, -um — of the Strait of Magellan region.
magnificus, -a, -um — magnificent, eminent, distingushed.
magnifolius, -a, -um — with large leaves.
majalis, -is, -e — flowering in May.
major, -or, -us, — larger, greater.
malacoides, -es, -es — soft, mucilaginous.
malacophyllus, -a, -um — soft-leaved.
maliformis, -is, -e — apple-shaped.
malvaceus, -a, -um — malva-like, mallow-like.
malvaeflorus, -a, -um — mallow-flowered.
mamillatus, -a, -um, mamillaris, -is, -e — having teat-shaped processes.
mammosus, -a, -um — with breasts or nipples.
mammulosus, -a, -um — with small nipples.
mandschuricus, -a, -um — of Manchuria.
mandushuricus, -a, -um — of Manchuria.
manicatus, -a, -um — manicate, long-sleeved, covered densely as with thick hairs so that the covering can be removed as such.
marcidus, -a, -um — faded.
margaritaceus, -a, -um — pearl-like.
margaritus, -a, -um — pearly, of pearls.
marginalis, -is, -e — marginal.
marginellus, -a, -um — somewhat margined.
marianus, -a, -um — of the Maryland region (U.S.A.).
marilandicus, -a, -um — of the Maryland region (U.S.A.).
maritimus, -a, -um — growing near the sea.
marmoratus, -a, -um — marbled, mottled.
maroccanus, -a, -um — of Morocco.
marylandicus, -a, -um — of the Maryland region (U.S.A.).
matronalis, -is, -e — pertaining to matrons.
mauritanicus, -a, -um — of Mauretania (N. Africa).
maximus, -a, -um — the largest.
mediopictus, -a, -um — pictured or stripped at the center.
medius, -a, -um — medium, intermediate.
megacarpus, -a, -um — large-fruited.
megalanthus, -a, -um — large-flowered.
megarrhizus, -a, -um — large-rooted.
megastachys, -ys, -ys, — large-spiked.
melancholicus, -a, -um — melancholy, hanging, or drooping.
melanocarpus, -a, -um — black-fruited.
melanococcus, -a, -um — black-berried.
melanotrichus, -a, -um — black-haired.
melantherus, -a, -um — with black anthers.
meleagris (noun) — a guinea-fowl.

melleus, -a, -um — pertaining to honey.
mellitus, -a, -um — honey, sweet.
meridionalis, -is, -e — southern.
metallicus, -a, -um — metallic (color or luster).
micans (pres. part.) — glittering, sparkling, mica-like.
micracanthus, -a, -um — small-spined.
micranthus, -a, -um — small-flowered.
microcarpus, -a, -um — small-fruited.
microcephalus, -a, -um — small-headed.
microlepis, -is, -is — small-scaled.
microphyllus, -a, -um — small-leaved.
miliaceus, -a, -um — pertaining to millet.
millefolius, -a, -um — many-leaved, thousand-leaved.
mimus, -a, -um — mimic.
minax, -ax, -ax — threatening, forbidding.
minimus, -a, -um — the smallest.
minor, -or, -us — smaller.
minutiflorus, -a, -um — minute-flowered.
mirabilis, -is, -e — wonderful, admirable.
mirus, -a, -um — wonderful, unusual.
mitis, -is, -e — mild, gentle.
modestus, -a, -um — modest.
moesiacus, -a, -um — of the Balkan region (ancient Moesians).
mollis, -is, -e — soft, soft-hairy.
mollissimus, -a, -um — very soft-hairy.
moluccanus, -a, -um — of the Moluccas (E. Indies).
moniliformis, -is, -e — necklace-shaped.
monocarpus, -a, -um — with one carpel.
monocephalus, -a, -um — with one head.
monococcus, -a, -um — one-berried.
monoicus, -a, -um — monoecious.
monopetalus, -a, -um — with one petal.
monophyllus, -a, -um — with one leaf.
monosepalus, -a, -um — with one sepal.
monospermus, -a, -um — one-seeded.
monspeliensis, -is, -e — living in caves and mountains.
monspessulanus, -a, -um — of Montpelier (France).
monstrosus, -a, -um — monstrous, abnormal.
montanus, -a, -um — of mountains.
montevidensis, -is, -e — of Montevideo (Uruguay).
monticolus, -a, -um — inhabiting mountains.
monumentalis, -is, -e — monumental.
mosaicus, -a, -um — parti-colored, as of mosaic.
moschatus, -a, -um — musky, musk-scented.
mucidus, -a, -um — moldy.
mucilaginosus, -a, -um — slimy.
mucronatus, -a, -um — with a short, small, abrupt tip.
multicaulis, -is, -e — with many stems.
multicostatus, -a, -um — many-ribbed.
multifidus, -a, -um — many-parted.
multiflorus, -a, -um — many-flowered.
multinervis, -is, -e — many-nerved.
multiradiatus, -a, -um — with numerous rays.
mundulus, -a, -um — trim, neat.
munitus, -a, -um — armed, fortified.
muralis, -is, -e — growing on walls.
muricatus, -a, -um — muricate, roughened by means of hard points.
murinus, -a, -um — mouse-colored.

musaicus, -a, -um — musa-like.
muscosus, -a, -um — mossy.
mutabilis, -is, -e — variable.
muticus, -a, -um — curtailed, blunt.
myriacanthus, -a, -um — myriad-spined.
myriophyllus, -a, -um — myriad-leaved.
myrmecophilus, -a, -um — ant-loving.

N

nanellus, -a, -um — very small or dwarf.
nanus, -a, -um — dwarf.
napiformis, -is, -e — turnip-shaped.
narbonensis, -is, -e — of Narbonne (ancient region of S. France).
natalensis, -is, -e — of Natal (S. Africa).
natans (pres. part.) — swimming or floating.
neapolitanus, -a, -um — Neapolitan, of Naples.
nebulosus, -a, -um — nebulous, clouded, indefinite.
neglectus, -a, -um — neglected.
nemoralis, -is, -e — growing in a wood.
nemorosus, -a, -um — growing in a wood or grove.
nemus (noun) — wood.
nephrolepis, -is, -is — kidney scaled.
neriifolius, -a, -um — nerium-leaved, oleander-leaved.
nervatus, -a, -um — nerved.
nervosus, -a, -um — nerved.
nictitans (pres. part.) — blinking, winking.
nidulus (noun) — small nest.
nidus (noun) — nest.
nigellus, -a, -um — blackish.
niger, -gra, -grum — black.
nigrescens (pres. part.) — blackish, becoming black.
nigricans (pres. part.) — blackish, becoming black.
nigrofructus, -a, -um — black-fruited.
nitens (pres. part.) — glittering, shining.
nitidus, -a, -um — glittering.
nivalis, -is, -e — snowy, pertaining to snow.
niveus, -a, -um — snow white.
nivosus, -a, -um — snowy, full of snow.
nobilior, -or, -us — more noble.
nobilis, -is, -e — noble.
noctiflorus, -a, -um — night-flowering.
nocturnus, -a, -um — of the night, night-blooming.
nodiflorus, -a, -um — with flowers at nodes.
nodosus, -a, -um — with nodes.
nodulosus, -a, -um — with small nodes.
nolitangere (verb) — do not touch, touch-me-not.
nonpinnatus, -a, -um — not pinnate.
nonscriptus, -a, -um — not described.
nootkatensis, -is, -e — of Nootka (Nootka Sound near Vancouver Island).
notatus, -a, -um — marked.
novae-angliae (noun) — of New England.
noveboracensis, -is, -e — of New York.
novi-belgi (noun) — of New Belgium (early name for New York).
nox (noun) — night.
nucifer, -era, -erum — nut bearing.
nudatus, -a, -um — nude, naked, exposed.
nudicaulis, -is, -e — naked-stemmed.
nudiflorus, -a, -um — naked-flowered.
nudus, -a, -um — naked, nude, exposed.
numidicus, -a, -um — of Numidia (ancient country of N. Africa).
nutans (pres. part.) — nodding.
nutkatensis, -is, -e — of Nootka (Nootka Sound near Vancouver Island).
nux (noun) — nut.
nyctagineus, -a, -um — night-blooming.

O

obcordatus, -a, -um — inversely cordate.
oblanceolatus, -a, -um — inversely lanceolate.
obliquus, -a, -um — oblique.
oblongatus, -a, -um — oblong.
oblongifolius, -a, -um — oblong-leaved.
obovatus, -a, -um — obovate, inverted ovate.
obtusatus, -a, -um — blunt, rounded.
obtusifolius, -a, -um — obtuse-leaved.
obtusus, -a, -um — obtuse, blunt, rounded.
occidens, -ens, -ens — west.
occidentalis, -is, -e — western.
ocellatus, -a, -um — like an eye.
ocellus (noun) — small eye.
ochraceus, -a, -um — ochre-colored
ochroleucus, -a, -um — yellowish-white.
octandrus, -a, -um — with eight stamens.
octoflorus, -a, -um — eight-flowered.
octopetalus, -a, -um — eight-petaled.
odessanus, -a, -um — of Odessa (Black Sea region).
odoratissimus, -a, -um — very fragrant.
odoratus, -a, -um — with an odor.
officinalis, -is, -e — official, medicinal, recognized in Pharmacopoeia.
officinarum (noun) — of the apothecaries.
oleaceus, -a, -um — oily.
oleifer, -era, -erum — oil-bearing.
oleraceus, -a, -um — herbaceous, oleraceous, garden herb used in cooking.
oliganthus, -a, -um — few-flowered.
oligocarpus, -a, -um — few-fruited.
oligospermus, -a, -um — few-seeded.
olitorius, -a, -um — pertaining to vegetable gardens or gardners.
olivaceus, -a, -um — olive green.
olympicus, -a, -um — of Olympus or Mt. Olympus (Greece).
opacus, -a, -um — opaque, shaded, not transparent.
operculatus, -a, -um — with a lid.
oppositiflorus, -a, -um — with opposite flowers.
oppositifolius, -a, -um — with opposite leaves.
opulifolius, -a, -um — with leaves like those of *Opulus*.
orbicularis, -is, -e — obicular.
orbiculatus, -a, -um — round or circular, disk-shaped.
orchiodes, -es, -es — orchid-like.
orchioides, -es, -es — orchid-like.
orientalis, -is, -e — eastern, oriental.
ornatissimus, -a, -um — very showy.
ornatus, -a, -um — adorned, ornate.
ornithocephalus, -a, -um — like a bird's head.
ornithopodus, ornithopus, -a, -um — like a bird's foot.
orthobotrys, -ys, -ys — straight-clustered.
orthocarpus, -a, -um — straight-fruited.
orthopterus, -a, -um — straight-winged.
orthosepalus, -a, -um — straight-sepaled.
ostrinus, -a, -um — purple.
ovalifolius, -a, -um — oval-leaved.
ovalis, -is, -e — oval.
ovatifolius, -a, -um — ovate-leaved.
ovatus, -a, -um — ovate.
ovinus, -a, -um — pertaining to sheep.
oxycanthus, -a, -um — sharp-spined.
oxygonus, -a, -um — sharp-angled, acute-angled.
oxyphyllus, -a, -um — sharp-leaved.

P

pabularius, -a, -um — of fodder or pasturage.
pachyanthus, -a, -um — thick-flowered.
pachycarpus, -a, -um — with thick pericarp.
pachyphlaeus, -a, -um — thick-barked.
pachyphyllus, -a, -um — thick-leaved.
pacificus, -a, -um — of the Pacific, or regions bordering the Pacific Ocean.
paleaceus, -a, -um — like chaff.
palestinus, -a, -um — of Palestine.
pallens, -ens, -ens — pale.
pallescens (pres. part.) — becoming pale.
palliatus, -a, -um — cloaked.
pallidiflorus, -a, -um — with pale flowers.
pallidus, -a, -um — pallid, pale.
palmatifidus, -a, -um — palmately divided.
palmatus, -a, -um — palmate.
paludosus, -a, -um — marshy.
palus (noun) — a marsh.
paluster, -tris, -tre — of marshes.
pampinus, -a, -um — tendril.
panduraeformis, -is, -e — violin-shaped.
paniculatus, -a, -um — in panicles.
paniculiger, -era, -erum — panicle-bearing.
panis (noun) — bread.
pannosus, -a, -um — ragged, tattered.
papaveraceus, -a, -um — poppy-like.
papilionaceus, -a, -um — like a butterfly.
papillosus, -a, -um — with small protuberances.
papposus, -a, -um — with a pappus.
papyraceus, -a, -um — papery.
papyrifer, -era, -erum — paper-bearing.
paradisiacus, -a, -um — of parks or gardens.
paradoxus, -a, -um — unusual, strange.
pardalinus, -a, -um — leopard-like, spotted.
pardalis (noun) — a panther.
partitus, -a, -um — parted.
parviflorus, -a, -um — small-flowered.
parvifolius, -a, -um — small-leaved.
parvulus, -a, -um — very small.
parvus, -a, -um — small.
pascuus, -a, -um — of pastures.
passerinus, -a, -um — pertaining to a sparrow.
pastoralis, -is, -e — pastoral.
patellaris, -is, -e — circular, disk-shaped.
patens (pres. part.) — spreading.
patiens, -ens, -ens — patient.
patulus, -a, -um — spreading.
pauciflorus, -a, -um — few-flowered.
paucifolius, -a, -um — few-leaved.
pauperculus, -a, -um — poor.
pavonicus, -a, -um — variegated.
pavoninus, -a, -um — peacock-like, variegated.
pectinatus, -a, -um — comblike, like the teeth in a comb.
pectinifer, -era, -erum — comb-bearing.
pectoralis, -is, -e — shaped like a breastbone.
pedatus, -a, -um — palmately divided or parted.
pedecarpus, -a, -um — with stalked fruit.
pedicularius, -a, -um — with a stalk.
peduncularis, -is, -e — with peduncles.
pedunculatus, -a, -um — with peduncles.
pedunculosus, -a, -um — with many peduncles.
pekinensis, -is, -e — of Peking, China.
pellucidus, -a, -um — with transparent dots.
peltatus, -a, -um — shield-shaped,

as a leaf attached by its under side rather than by its margin.

peltifolius, -a, -um — peltate-leaved.

Penduliflorus, -a, -um — with pendulous flowers.

pendulinus, -a, -um — somewhat pendulous.

pendulus, -a, -um — pendulous, hanging.

penicillatus, -a, -um — hair-penciled.

pennatifidus, -a, -um — pennatifid.

pennatus, -a, -um — pinnate, feathered.

pensilis, -is, -e — pensile, hanging.

pentalophus, -a, -um — five-winged or five-tufted.

pentandrus, -a, -um — with five stamens.

pentasepalus, -a, -um — with five sepals.

perbellus, -a, -um — very beautiful.

peregrinus, -a, -um — foreign, exotic.

perennans, -ans, -ans — perennial.

perennis, -is, -e — perennial.

perfoliatus, -a, -um — with stem passing through a leaf.

perforatus, -a, -um — with holes.

perfosus, -a, -um — perfoliate.

pergracilis, -is, -e — very slender.

perpusillus, -a, -um — very small.

persicarius, -a, -um — like a peach.

persicus, -a, -um — of Persia, like a peach.

persistens. -ens, -ens — persistent.

perspicuus, -a, -um — transparent.

perulatus, -a, -um — wallet-like or pocket-like.

pervenustus, -a, -um — very beautiful.

perviridis, -is, -e — very green, deep green.

petiolaris, -is, -e — pertaining to the petiole.

petiolatus, -a, -um — with petioles.

petiolus, -a, -um — with petioles.

petraeus, -a, -um — stony.

petrocallis (noun) — rock beauty.

phaeocarpus, -a, -um — dark-fruited.

philadelphicus, -a, -um — of the Philadelphia (Penn., U.S.A.) region.

phleoides, -es, -es — resembling timothy.

phlogiflorus, -a, -um — flame-flowered, phlox-flowered.

pleniflorus, -a, -um — double-flowered.

pleurostachys, -ys, -ys — side-spiked.

plicatus, -a, -um — folded.

plumarius, -a, -um — with plumes or feathers.

plumosus, -a, -um — feathery.

podocarpus, -a, -um — with stalked fruits.

podophyllus, -a, -um — with stalked leaves.

poeticus, -a, -um — of or pertaining to poets.

politus, -a, -um — polished.

polyacanthus, -a, -um — many-spined.

polyanthus, -a, -um — many-flowered.

polycarpus, -a, -um — many-fruited.

polycephalus, -a, -um — many-headed.

polylepis, -is, -is — with many scales.

polymorphus, -a, -um — of many forms, variable.

polystachys, -ys, -ys — with many spikes.

polystictus, -a, -um — many-dotted.

pomeridianus, -a, -um — afternoon.

pomifer, -era, -erum — pome-bearing.

pomiformis, -is, -e — shaped like a pome.

ponderosus, -a, -um — heavy, ponderous.

ponticus, -a, -um — of Ponticus (in Asia Minor).

populifolius, -a, -um — poplar-leaved.

populneus, -a, -um — pertaining to poplars.
porcinus, -a, -um — pertaining to swine.
porphyreticus, -a, -um — purple.
porrifolius, -a, -um — porrum-leaved, leek-leaved.
portoricensis, -is, -e — of Puerto Rico.
portulaceus, -a, -um — portulaca-like, purslane-like.
praealtus, -a, -um — very tall.
praecox, -ox, -ox — precocious, premature, very early.
praemorsus, -a, -um — bitten at the end.
praestans (pres. part.) — distinguished, excelling.
praetextus, -a, -um — bordered.
prasinatus, -a, -um — greenish.
prasinus, -a, -um — grass green.
pratensis, -is, -e — growing in meadows.
pravissimus, -a, -um — very crooked.
precatorius, -a, -um — praying, prayerful.
primulinus, -a, -um — primrose-yellow.
princeps, -ceps, -ceps — princely, first.
prismatocarpus, -a, -um — prism-fruited.
proboscideus, -a, -um — proboscis-like.
procerus, -a, -um — tall.
procumbens (pres. part.) — lying on the ground.
procurrens (pres. part.) — extending.
profusus, -a, -um — profuse.
prolifer, -era, -erum — prolific, fruitful.
prolificus, -a, -um — prolific, fruitful.
propendens (pres. part.) — hanging down.
prostratus, -a, -um — lying on the ground.
provincialis, -is, -e — provincial.
pruinosus, -a, -um — hoarfrost; covered with whitish dust or bloom.
prunifolius, -a, -um — plum-leaved.
pruriens (pres. part.) — itching.
pseudacacius (noun) — false acacia.
pseudonarcissus (noun) — false narcissus.
pseudoplatanus (noun) — false plane tree.
psilostachyus, -a, -um — naked-spiked.
psilostemon, -on, -on — smooth-stamened, naked-stamened.
psycodes, -es, -es — fragrant.
pteranthus, -a, -um — with winged flowers.
pterocarpus, -a, -um — with winged fruit.
pubens, -ens, -ens — downy.
pubescens, -ens, -ens — downy, covered with soft hairs.
pubiflorus, -a, -um — downy-flowered.
pubiger, -era, -erum — down-bearing.
pudicus, -a, -um — bashful, retiring, shrinking, modest, chaste.
pugioniformis, -is, -e — dagger-shaped.
pulchellus, -a, -um — pretty, beautiful.
pulcherrimus, -a, -um — very handsome, very beautiful.
pulverulentus, -a, -um — powdered, dust-covered.
pumilus, -a, -um — dwarf, small.
punctatus, -a, -um — dotted.
pungens, -ens, -ens — prickly, piercing, sharp-pointed.
punicans, -ans, -ans — reddish.
puniceus, -a, -um — reddish-purple, crimson.
purgans (pres. part.) — purging.
purpurascens (pres. part.) — purplish, becoming purple.
purpureus, -a, -um — purple.
pusillus, -a, -um — dwarf, small.
pycnacanthus, -a, -um — densely spiny.

pycnanthus, -a, -um — densely flowered.
pycnocephalus, -a, -um — thick-headed.
pygmaeus, -a, -um — dwarf.
pyramidalis, -is, -e — pyramidal.
pyrenaicus, -a, -um — of the Pyrenees.
pyriformis, -is, -e — shaped like a pear.

Q

quadrangularis, -is, -e, quadrangulatus, -a, -um — four-angled.
quadriflorus, -a, -um — with four flowers.
quadrifolius, -a, -um — with four leaves.
quercifolius, -a, -um — oak-leaved.
quinqueflorus, -a, -um — with five flowers.
quinquefolius, -a, -um — with five leaves.

R

racemosus, -a, -um — in racemes, resembling racemes.
radiatus, -a, -um — radiate, rayed.
radicalus, -a, -um — from the root.
radicans (pres. part.) — rooting.
ramosissimus, -a, -um — many-branched.
ramosus, -a, -um — branched.
rapiformis, -is, -e — turnip-shaped.
reclinatus, -a, -um — reclining, bent back.
recurvus, -a, -um — curved back.
redivivus, -a, -um — restored, brought to life.
reflexus, -a, -um — bent back.
refractus, -a, -um — curved, broken.
regalis, -is, -e — royal, regal.
regina (noun) — queen.
regius, -a, -um — royal, magnificent, kingly.
religiosus, -a, -um — used for religious purposes, venerated, sacred.
reniformis, -is, -e — kidney-shaped.
repens (pres. part.) — creeping.
reptans (pres. part.) — creeping.
resinosus, -a, -um — resinous.
resupinatus, -a, -um — inverted.
reticulatus, -a, -um — netlike.
retortus, a, -um — bent back.
retroflexus, -a, -um — reflexed.
retrofractis, -is, -e — broken or bent backwards.
retusus, -a, -um — retuse, notched slightly at a rounded apex.
revolutus, -a, -um — revolute, rolled backwards.
rhizophyllus, -a, -um — root-leaved, leaves rooting.
rhombifolius, -a, -um — with rhombic leaves.
rhomboidalis, -is, -e — with rhombic outline.
rhytidophyllus, -a, -um — wrinkle-leaved.
rigens, -ens, -ens — rigid, stiff.
rigidisetae (noun) — stiff bristles.
rigidus, -a, -um — stiff.
ringens (pres. part.) — gaping, open-mouthed.
riparius, -a, -um — growing on side of river.
rivalis, -is, -e — growing on brookside.
rivularis, -is, -e — growing on brookside.
roborosus, -a, -um — strong.
robustus, -a, -um — robust, stout.
roridus, -a, -um — dewy, covered with particles which resemble dew.
rosaeflorus, -a, -um — rose-flowered.
roseus, -a, -um — rose-colored.
rostratus, -a, -um — rostrate, beaked.
rosulatus, -a, -um — like a rosette.
rotundifolius, -a, -um — round-leaved.
rotundus, -a, -um — round.
rubellus, -a, -um — reddish.
rubens, -ens, -ens — red, ruddy.
ruberrimus, -a, -um — very red.
rubescens (pres. part.) becoming red.

rubicundus, -a, -um — reddish.
rubidus, -a, -um — dark red.
rubiginosus, -a, -um — brownish-red.
ruber, -ra, -rum — red.
rudis, -is, -e — wild, not tilled.
rufidulus, -a, -um — somewhat rufid, reddish.
rufus, -a, -um — fox red.
rugosus, -a, -um — wrinkled.
rugulosus, -a, -um — somewhat rugose.
rupestris, -is, -e — growing on rocks.
rupicolus, -a, -um — growing on rocks.
russus, -a, -um — red.
rusticanus, -a, -um — rustic, pertaining to the country.

S

saccatus, -a, -um — saccate, bag-like.
saccharatus, -a, -um — containing sugar.
saccharinus, -a, -um — saccharine, of sugar.
saccharoides, -es, -es — sweet, like sugar.
saccharum (noun) — sugar.
saccus (noun) sac.
sachalinensis, -is, -e — of Saghalin Island (N. Japan).
sacrorum (noun) — of sacred places.
sagittifolius, -a, -um — arrow-leaved.
sagittalis, -is, -e -- arrow-shaped.
sagittatus, -a, -um — arrow-shaped.
salicifolius, -a, -um — willow-leaved.
salicis, -is, -e — of the willow.
salignus, -a, -um — willow-like, of the willow.
salinus, -a, -um — salty.
salsuginosus, -a, -um — fond of salt marshes.
sanctus, -a, -um — holy.
sanguinalis, -is, -e — blood-colored.
sanguineus, -a, -um — blood red.
sanguis (noun) — blood.
sapidus, -a, -um — tasteful, savory, pleasing to taste.
sapientum (noun) — of the wise men or authors.
saponarius, a, -um — soapy.
sarmentosus, -a, -um — sarmentose, bearing runners.
sativus, -a, -um — cultivated.
saxatilis, -is, -e — growing on rocks.
saxifragus, -a, -um — stone-breaking.
scaber, -bra, -brum — rough, not smooth.
scaberrimus, -a, -um — very rough.
scandens (pres. part.) — climbing.
scaposus, -a, um — with scapes.
scariosus, -a, -um — shriveled, thin and not green.
scarletinus, -a, -um — scarlet.
sceleratus, -a, -um — pernicious.
sceptrum (noun) — scepter.
schistaceus, -a, -um — slate gray.
schistosus, -a, -um — schistose, slaty as to tint.
schizopetalus, -a, -um — with cut petals.
scilloides, -es, -es — like squill.
scissus, -a, -um — split.
sciureus, -a, -um — resembling squirrel's tail.
sclerophyllus, -a, -um — hard-leaved.
scoparius, -a, -um — broomlike.
scopulinus, -a, -um — rocklike.
scopulorum (noun) — of the rocks.
scorpioides, -es, -es — scorpion-like.
scutatus, -a, -um — buckler-shaped, like a small shield.
scutellatus, -a, -um — shield-shaped.
sebifer, -era, -erum — tallow-bearing.
sebosus, -a, -um — full of tallow or grease.
secalinus, -a, -um — like rye.
secundatus, -a, -um — secund, on the side.

secundiflorus, -a, -um — one-sided, secund-flowered.
secundus, -a, -um — secund, on the side.
segetum (noun) — of corn fields.
semidecandrus, -a, -um — with five stamens, half-of-ten stamens.
semiglobosus, -a, -um — half-globose.
semperflorens (pres. part.) — ever-blooming.
sempervirens, -ens, -ens — evergreen.
sempervivus, -a, -um — ever-living.
senescens (pres. part.) — becoming old or gray.
senilis, -is, -e — senile, old, white-haired.
sensibilis, -is, -e — sensible.
sensitivus, -a, -um — sensitive.
sepiarius, -a, -um — of or pertaining to hedges.
saepium (noun) — of hedges or fences.
septentrionalis, -is, -e — northern.
sericeus, -a, -um — silken.
sericifer, -era, -erum — silk-bearing.
sericofer, -era, -erum — silk-bearing.
serotinus, -a, -um — late-flowering, fall.
serpens (pres. part.) — creeping, crawling.
serpentinus, -a, -um — of snakes, serpentine, looping or waving.
serratifolius, -a, -um — with serrate leaves.
serratus, -a, -um — serrate, saw-toothed.
serrulatus, -a, -um — finely serrate or saw-toothed.
sesquipedalis, -is, -e — one foot and a half long or high.
sessiliflorus, -a, -um — with stemless flowers.
sessilifolius, -a, -um — with stemless leaves.
sessilis, -is, -e — stemless.
sessilispicatus, -a, -um — with a spike without pedicel.
setaceus, -a, -um — bristle-like.
setifolius, -a, -um — bristle-leaved.
setiger, -era, -erum — bristly, bristle-bearing.
setosus, -a, -um — with bristles.
sibericus, -a, -um — of Siberia.
siculus, -a, -um — of Sicily.
signatus, -a, -um — marked, designated, attested.
silvaticus, -a, -um — pertaining to woods.
silvestris, -is, -e — pertaining to woods or forest.
similis, -is, -e — similar, like.
simplex, -ex, -ex — simple, unbranched.
simplicifolius, -a, -um — simple-leaved.
sinensis, -is, -e — Chinese, of China.
sinuatus, -a, -um — sinuate, wavy-margined.
siphiliticus, -a, -um — syphilitic.
smaragdinus, -a, -um — emerald green.
sobolifer, -era, -erum — bearing creeping, rooting stems or shoots, sucker-bearing.
somnifer, -era, -erum — sleep-bearing, sleep-producing.
somnus (noun) — sleep.
sordidus, -a, -um — dirty.
sparsiflorus, -a, -um — with scattered flowers.
sparsus, -a, -um — scattered.
sparteus, -a, -um — spearlike.
spathulatus, -a, -um — with a spathe.
spatiosus, -a, -um — spacious, wide.
speciosus, -a, -um — beautiful, showy, good-looking.
spectabilis, -is, -e — spectacular, remarkable.
speculatus, -a, -um — shining, as if with mirrors.
sphaerocarpus, -a, -um — spherical-fruited.
sphaerocephalus, -a, -um — spherical-headed.
spicatus, a, -um — with spikes.
spiciformis, -is, -e — spike-shaped.

spinescens (pres. part.) — becoming spiny.
spinifer, -era, -erum — bearing spines.
spinosus, -a, -um — with spines or thorns.
spinulosus, -a, -um — with small spines.
spiralis, -is, -e — spiral.
splendens (pres. part.) — glittering, splendid.
splendidus, -a, -um — splendid.
spongiosus, -a, -um — spongy.
spontaneus, -a, -um — spontaneous.
spurius, -a, -um — spurious, false, bastard.
squamatus, -a, -um — squamate, with small scalelike leaves or bracts.
squamiger, -a, -um — scale-bearing.
squamosus, -a, -um — squamate, full of scales.
squarrosus, -a, -um — squarrose, with parts spreading or even recurved at ends.
stagninus, -a, -um — growing in a marsh.
stamineus, -a, -um — with prominent stamens.
stans (pres. part.) — erect, standing, upright.
stellatus, -a, -um — with stars, starlike.
stellipilus, -a, -um — with stellate hairs.
stellulatus, -a, -um — somewhat stellate.
stenocephalus, -a, -um — narrow-headed.
stenophyllus, -a, -um — narrow-leaved.
sterilis, -is, -e — sterile, infertile.
stipulatus, -a, -um — having stipules.
stolonifer, -era, -erum — with stolons or runners that take root.
stramineofructus, -a, -um — with straw-colored fruit.
stramineus, -a, -um — straw-colored.
streptocarpus, -a, -um — twisted-fruited.
streptophyllus, -a, -um — twisted-leaved.
striatus, -a, -um — with parallel channels or lines, striate.
strictus, -a, -um — stiff, rigid, upright, erect.
strigosus, -a, -um — with stiff bristles.
strobilus, -a, -um — cone.
strumosus, -a, -um — strumous, having cushion-like swellings.
stylosus, -a, -um — with style or styles prominent.
styracifluus, -a, -um—flowing with storax or gum.
suaveolens, -ens, -ens — sweet-smelling.
suavis, -is, -e — sweet, agreeable.
subalpinus, -a, -um — below the alpine.
subcordatus, -a, -um — almost cordate.
suberosus, -a, -um — corky.
submersus, -a, -um — submerged.
subovatus, -a, -um — almost ovate.
subterraneus, -a, -um — under the ground.
subulatus, -a, -um — awl-shaped.
succidus, -a, -um — sappy.
succulentus, -a, -um — succulent, fleshy.
sudanensis, -is, -e — of the Sudan.
suecius, -a, -um — of Sweden.
suffruticosus, -a, -um — shrubby.
sulcatus, -a, -um — furrowed.
sulfureus, -a, -um—sulfur-colored.
superbus, -a, -um — superb, proud.
supinus, -a, -um — prostrate.
suprafoliaceus, -a, -um — above the leaf.
suspensus, -a, -um — suspended, hung.
sylvaticus, -a, -um — sylvan, forest-loving.
sylvestris, -is, -e — growing in woods (forest).
symphoricarpus, -a, -um — with fruits together.
syphiliticus, -a, -um — syphilitic.

T

tabacinus, -a, -um — like tobacco.
tabulaeformis, -is, -e — table-shaped.
tamariscifolius, -a, -um — tamarisk-leaved.
tanacetifolius, -a, -um — tansy-leaved.
tardiflorus, -a, -um — late-flowering.
tardus, -a, -um — late.
tartareus, -a, -um — with loose or rough crumbling surface.
tataricus, -a, -um — of Tartary (cent. Asia).
taxifolius, -a, -um — yew-leaved.
tectorus, -a, -um — of roofs, or houses.
tectus, -a, -um — concealed, covered.
temulentus, -a, -um — intoxicating.
tenarius, -a, -um — slender.
tenax, -ax, -ax — tenacious, strong.
tenebrosus, -a, -um — of dark or shaded places.
tenellus, -a, -um—flexible, slender.
tener, -a, -um — slender, tender, soft.
tenuiflorus, -a, -um—slender flowers.
tenuifolius, -a, -um — slender-leaved.
tenuior, -or, -us — very slender.
tenuis, -is, -e — slender, thin.
terebinthinus, -a, -um — of turpentine.
teres, -es, -es — terete, circular in cross section.
tereticornis, -is, -e — with terete or cylindrical horns.
teretifolius, -a, -um — terete-leaved.
terminalis, -is, -e — terminal, at the end of a stem or branch.
ternatus, -a, -um — arranged in threes.
ternifolius, -a, -um — with leaves in threes.
terrestris, -is, -e — of the earth.
tessellatus, -a, -um — tessellate, laid off in squares or in dicelike pattern.
testaceus, -a, -um — light brown, brick-colored.
testiculatus, -a, -um — testiculated, testicled.
testudinarius, -a, -um — like a tortoise-shell.
tetragonolobus, -a, -um — with a four-angled pod.
tetrapterus, -a, -um—four-winged.
tetrastachyus, -a, -um — with four spikes.
textilis, -is, -e — textile, woven.
thelypteroides, -es, -es — thelypterum-like.
thermalis, -is, -e — warm, of warm springs.
thuyoides, -es, -es — like *Thuja* (*Thuya*) or arborvitae.
thyoides, -es, -es — See thuyoides.
thyrsiflorus, -a, -um — flowers borne in a thyrsus.
thyrsoideus, -a, -um — thyrse-like.
tigrinus, -a, -um — tiger-striped.
tiliaceus, -a, -um — Tilia-like (like linden or basswood).
tinctorius, -a, -um — a dyer.
tinctus, -a, -um — dyed.
tomentosus, -a, -um—felty, downy.
torminalis, -is, -e — useful against colic.
torosus, -a, -um — cylindrical with contractions at intervals.
tortilis, -is, -e — twisted.
tortuosus, -a, -um — much twisted.
torulosus, -a, -um — somewhat torose or contracted at intervals.
transalpinus, -a, -um—transalpine.
tremuloides, -es, -es — trembling, quaking.
tremulus, -a, -um — quivering, trembling.
triacanthus, -a, -um — three-spined.
triangularis, -is, -e — triangular.
tricephalus, -a, -um — three-headed.
trichocarpus, -a, -um — hairy-fruited.
trichodes, -es, -es — pilose.

trichosanthus, -a, -um — with hairy flowers.
trichospermus, -a, -um — hairy-seeded.
trichotomus, -a, -um — three-branched or three-forked.
tricoccus, -a, -um — three-lobed.
tricolor, -or, -or — three-colored.
tricornis, -is, -e — three-horned.
tridens, -ens, -ens — with three teeth.
trifidus, -a, -um — three-cleft or three-parted.
trifoliatus, -a, -um — three-leaved.
trifurcatus, -a, -um — three-forked.
trilobatus, -a, -um — three-lobed.
trimestris, -is, -e—of three months, as lasting that time or maturing in it.
trinervis, -is, -e — three-nerved.
trinotatus, -a, -um — three-marked or three-spotted.
tripartitus, -a, -um — three-parted.
tripetalus, -a, -um — three-petaled.
triphyllus, -a, -um — three-leaved.
tripterius, -a, -um — three-winged.
trispermus, -a, -um — three-seeded.
tristachyus, -a, -um—three-spiked.
tristis, -is, -e — sad, dull, bitter.
trivialis, -is, -e — common, ordinary, found everywhere.
truncatus, -a, -um — ending abruptly.
tubaeformis, -is, -e — trumpet-shaped.
tubatus, -a, -um—trumpet-shaped.
tuberosus, -a, -um — with tubers.
tubiflorus, -a, -um — trumpet-flowered.
tubulosus, -a, -um — tubular, with tubes.
tumidus, -a, -um — swollen.
tunicatus, -a, -um — with concentric coats.
turbinatus, -a, -um — top-shaped.
turgidus, -a, -um — turgid, inflated, full.
typhinus, -a, -um — pertaining to fever.
typicus, -a, -um — typical.

U

uliginosus, -a, -um — of wet or marshy places.
umbellatus, -a, -um — with umbels.
umbellifer, -a, -um — bearing umbels.
umbraculifer, -a, -um — bearing an umbrella.
umbrosus, -a, -um — shaded or shade-loving.
uncinatus, -a, -um — hooked at the point, with hooks.
undatus, -a, -um — wavy.
undulatus, -a, -um — with waves.
unguicularis, -is, -e — with claws, tapered to a petiole-like base.
unguipetalus, -a, -um — with clawed petals.
unicornis, -is, -e — one-horned.
uniflorus, -a, -um — one-flowered.
unilabiatus, -a, -um — one-lipped.
unilateralis, -is, -e — one-sided.
unioloides, -es, -es — uniola-like.
univittatus, -a, -um — with one longitudinal stripe.
urceolatus, -a, -um—like a pitcher.
urceolus (noun)— small pitcher.
urens (pres. part.) — burning, stinging.
urniger, -a, -um — pitcher-bearing.
urostachyus, -a, -um — tail-spiked.
ursinus, -a, -um — pertaining to bears, northern (under the Great Bear).
urticaefolius, urticifolius, -a, -um— nettle-leaved.
urticoides, -es, -es — nettle-like.
usitatissimus, -a, -um — most commonly used.
usneoides, -es, -es — usnea-like.
utilis, -is, -e — useful.
utricularis, -is, -e — bladder-shaped.
utriculus (noun) — sac.
uvaeformis, -is, -e — like a raceme.

V

vacillans (pres. part.) — swaying.
vagans (pres. part.) — vagrant, wandering.
vaginaeflorus, -a, -um — with flowers in sheath.
vaginans (pres. part.) — sheathing, wrapping around.
vaginatus, -a, -um—sheathed, with a sheath.
valdivianus, -a, -um — of Valdivia (Chile).
valdiviensis, -is, -e — of the region of Valdivia (Chile).
variabilis, -is, -e — variable, of many forms.
varicosus, -a, -um — varicose, with veins or filaments dilated.
variegatus, -a, -um — variegated.
variifolius, -a, -um — variable-leaved.
velox, -ox, -ox — rapidly growing.
velutinus, -a, -um — velvety.
venenatus, -a, -um — poisonous.
venenosus, -a, -um — venomous, poisonous.
venosus, -a, -um — with veins.
ventralis, -is, -e — facing the center of a flower, opposite of dorsal.
ventricosus, -a, -um — ventricose, having a swelling or inflation on one side.
venustus, -a, -um — beautiful, handsome, charming.
vermiculatus, -a, -um — wormlike.
vernalis, -is, -e — vernal, flowering in the spring, of spring.
vernicosus, -a, -um — varnished.
vernus, -a, -um — of spring, vernal.
verrucosus, -a, -um — verrucose, warted.
verruculosus, -a, -um — very warty or verrucose.
versatilis, -is, -e — movable, swinging freely.
versicolor, -or, -or — variously colored, as of one color blending into another, of changing color.
versiflorus, -a, -um — with different flowers.
versutus, -a, -um — versatile.
verticillatus, -a, -um — whorled, arranged in a circle about the stem.
vertucosus, -a, -um — with warts, verrucose.
verus, -a, -um — the true, genuine, standard.
vesciculosus, -a, -um — with little bladders.
vescus, -a, -um — weak, thin, feeble.
vesicarius, -a, -um — bladder-shaped.
vesicularis, -is, -e — with little bladders or blisters.
vespertinus, -a, -um — blooming in the evening.
vestitus, -a, -um — covered or clothed with hairs, pubescent.
vexillarius, -a, -um — of the standard petal (pea flower), with a standard.
villosissimus, -a, -um — very hairy.
villosus, -a, -um — covered with soft hairs.
viminalis, -is, -e — of osiers, of basket willows.
vinifer, -a, -um — wine-bearing.
violaceus, -a, -um — violet-colored.
virens (pres. part.) — greening.
virescens (pres. part.) — becoming green.
virgatus, -a, -um — twiggy.
virginalis, -is, -e — pertaining to young women.
viridiflorus, -a, -um — with green flowers.
viridis, -is, -e — green.
viridissimus, -a, -um — very green.
viridulus, -a, -um — greenish.
virosus, -a, -um — virulent, poisonous, fetid.
virulentus, -a, -um — virulent, poisonous, fetid.
viscarius, -a, -um — viscid, like mistletoe.
viscidulus, -a, -um — somewhat sticky.
viscidus, -a, -um — viscid, sticky.
viscosus, -a, -um — viscid, sticky.

vitaceus, -a, -um — vitis-like, like the grape.
vitellinus, -a, -um — dull yellow approaching red.
vitifolius, -a, -um — vitis-leaved, grape-leaved.
vittatus, -a, -um — striped.
vittiger, -a, -um — bearing stripes.
viviparus, -a, -um — freely producing asexual propagating parts, as bulbets in the inflorescence.
vixcordatus, -a, -um — nearly heart-shaped.
volatilis, -is, -e — volatile.
volubilis, -is, -e — twining.
vomicus, -a, -um — emetic.
vomitorius, -a, -um — emetic.
vulcanicus, -a, -um — of Vulcan or of a volcano, volcanic.
vulgaris, -is, -e — vulgar, common, usual.
vulnerarius, -a, -um — useful for wounds.
vulpes, -pes, -pes — fox.
vulpinus, -a, -um — pertaining to the fox.

X

xanthacanthus, -a, -um — yellow-spined.
xanthinus, -a, -um — yellow.
xanthocarpus, -a, -um — yellow-fruited.
xanthoneurus, -a, -um — yellow-nerved.
xanthorrhizus, -a, -um — yellow-rooted.

Y

yedoensis, -is, -e — of Yedo or Yeddo (Japan).
yunnanensis, -is, -e — of province of Yunnan (China).

Z

zebrinus, -a, -um — zebra-striped.
zeylanicus, -a, -um — Ceylonian, of Ceylon.
zibethinus, -a, -um — like the civet-cat, malodorous.
zizanioides, -es, -es — zizania-like.
zonalis, -is, -e — zonal.
zonatus, -a, -um — zonal, zoned, banded.

Some Greek and Latin Components of Scientific Words

These components are either Greek or Latin, or are derived from Greek or Latin. The prefixes are followed by a hyphen and the suffixes are preceded by a hyphen, while the other combining forms are without hyphens. In the construction of new words, Greek should be combined with Greek and Latin with Latin, otherwise a hybrid word is produced which is not considered good practice. In the joining of Latin stems, the letter **- i -** is inserted when a vowel is required, while in Greek the letter **- o -** is used.

The columns below from left to right are as follows:

(1) Greek and Latin word components
(2) Prefixes, suffixes and parts of speech indicated in English
(3) Meaning of respective word components
(4) Botanical name or term in either Latin, Greek, or English in which component is used.

Component	Type	Meaning	Example
A			
a- L.	pref.	from, away	averse
a-, an- Gr.	pref.	without, not	amorphous
ab- L.	pref.	See *a-*.	abaxial
abr Gr.	adj.	soft, delicate, dainty	*Abronia*
abund L.	adj.	abounding, full	*florabundus*
ac Gr., L.	n.	a point, a needle, a thorn	acicular
acanth Gr.	n.	a thorn, a prickle	*Pyracantha*
-aceae L.	suf.	a suffix used to denote plant family names	*Rosaceae*

Component	Type	Meaning	Example
-aceus L.	suf.	of or pertaining to, resembling	rosaceous
achyr Gr.	n.	chaff, husks	*Amphiachyris*
acr Gr.	adj.	at the top, apex or summit	acroblastic
acr L.	adj.	sharp	*acrimonia*
actin Gr.	n.	radiated, rayed	actinomorphic
-acus L.	suf.	of or belonging to	*aurantiacus*
acut L.	adj.	sharp, pointed	*acutifolius*
ad- L.	pref.	towards (the "d" is usually assimilated to a following consonant, as in aggregate, acclimate, etc.)	adaxial
-ad Gr.	suf.	daughter of	heliad
adelph Gr.	n.	brother	monadelphous
aden Gr.	n.	a gland	*Zygadenus*
aer Gr.	n.	air, atmosphere	*aeroides*, *anaerobic*
agath Gr.	adj.	good, excellent	agathophyllum
agn Gr.	adj.	pure, innocent	*Elaeagnus*
agr Gr.	n.	a field	*Agropyron*
agrost Gr.	n.	grass	*Eragrostis*
ai Gr.	adj.	always, ever	aiophyllous
-ales L.	suf.	used to denote plant orders	*Rosales*
-alis L.	suf.	pertaining to	*sanguinalis*
all Gr.	adj.	other, different, strange	allogamy
alopec Gr.	n.	a fox	*Alopecurus*
altern L.	adj.	in alternation	*alternifolius*
alt L.	adj.	high	altitude
alve L.	n.	hollow, cavity	alveolate
amm Gr.	n.	sand	ammophilous
ampel Gr.	n.	a vine	*Ampelopsis*
amphi- Gr.	pref.	on both sides of, about	*amphibius*
amplex L.	adj.	embracing, encircling, clasping	*amplexicaulis*
ana- Gr.	pref.	upward, throughout, backward	*anacamptis*
andr Gr.	n.	man, male	*Andropogon*
anem Gr.	n.	wind	*anemophilous*
angi Gr.	n.	a vessel, a container	angiosperm
angust L.	adj.	small, narrow	*angustifolius*
anis Gr.	adj.	unequal, dissimilar	anisopetalous
ante- L.	pref.	before, in front of	antechamber
anth Gr.	n.	a flower	*polyanthes*
anthrop Gr.	n.	man, human being	*anthropophilous*
anti- Gr.	pref.	against, opposite to	antidromy
annu L.	n.	a year	annual
anul L.	n.	a ring	*anularis*
-anus L.	suf.	of or belonging to	*floridanus*

Component	Type	Meaning	Example
ap-, aph- Gr.	pref.	from, away from, separate	apocarpous
apic L.	n.	summit, tip	apiculate
aqu L.	n.	water	*aquaticus*
arachn Gr.	n.	a spider	*arachnifer*
arbor L.	n.	tree	arboretum
arc L.	n.	a bow, bowlike	arcuate
arch- Gr.	pref.	beginning, first	archegonium
aren L.	n.	sand, a sandy place	*arenicola*
argyr Gr.	n.	silver	argyroneurous
-aris L.	suf.	of or pertaining to	*axilaris*
arist Gr.	adj.	noblest, best	*Aristolochia*
-arium L.	suf.	the place for a thing	herbarium
arrhen Gr.	adj.	male	*Diarrhena*
articul L.	adj.	a joint	articulate
arundo, arundin L.	n.	a reed, cane	*Arundinaria*
arv L.	n.	a field	*arvensis*
asper L.	adj.	rough	*Asperula*
aster Gr.	n.	a star	*asteroides*
atr L.	adj.	dark	*atrovirens*
-atus L.	suf.	provided with, made	ciliate
aur L.	n.	ear	auriculate
aur L.	adj.	golden	*auricomus*
austr L.	adj.	the south wind, southern	austral
aut Gr.	n.	self	autogamy

B

Component	Type	Meaning	Example
bapt Gr.	n.	dyed, bright-colored	*Baptisia*
barb L.	n.	beard	barbellate
bi Gr.	n.	life	biology
bi- L.	pref.	two, twice, double	bidentate
blast Gr.	n.	primitive germ, bud, sprout	blastochore
blephar Gr.	n.	eyelid, eyelash	*Blepharoneuron*
bol Gr.	v.	to throw	*Sporobolus*
bore Gr.	adj.	north	*borealis*
botry Gr.	n.	a cluster, a bunch of grapes	*Botrychium*
brachi L.	n.	the upper part of the arm	brachionate
brachy Gr.	adj.	short	brachypodous
brady Gr.	adj.	slow, heavy	*bradycarpus*
briz Gr.	adj.	nodding, sleepy	*Briza*
brom Gr.	n.	food, oats	*Bromus*
brot Gr.	n.	mortal, man	brotochore

Component	Type	Meaning	Example
C			
cact Gr.	n.	a prickly plant	*Cactaceae*
cad L.	adj.	falling	caducous
cal Gr.	adj.	beautiful	*Calochortus*
calam L.	adj.	a reed	*Calamagrostis*
calc L.	n.	lime	*calcicola*
calce L.	n.	a shoe, a slipper	*Calceolaria*
call L.	n.	a hard or thick skin	callus
call Gr.	adj.	beautiful	*Callicarpum*
calyc Gr.	n.	a covering, as that of a bud	*Calycanthus*
calyptr Gr.	n.	cover, lid	calyptrogen
campan L.	n.	a bell	campanulate
campt Gr.	v.	to bend, to curve	*Camptosorus*
campyl Gr.	adj.	bent, curved	campylotropous
can L.	n.	dog	*caninus*
can L.	n.	white, hoary	canescent
cand L.	adj.	white	*candidus*
capill L.	n.	hair	capillary
capit L.	n.	head	capitate
capreol L.	n.	a tendril, a support	capreolate
carn L.	n.	flesh	carneous
carp Gr.	n.	fruit	pericarp
carph Gr.	n.	twig, chaff, straw	hemicarphous
cartilag L.	n.	cartilage, gristle	cartilaginous
cary Gr.	n.	a nut	*caryopsis*
caud L.	n.	a tail	caudate
caul L.	n.	a stem, a stalk	*albicaulis*
cephal Gr.	n.	a head	*Cephalanthus*
cerat Gr.	n.	a horn	*Ceratophyllum*
cerc Gr.	n.	shuttle, peg	cercidium
cerc Gr.	n.	a tail	*Cercocarpus*
cero L.	adj.	waxen, of wax	*Ceropegia*
cern L.	adj.	turned downward, stooping	*cernuus*
chaer Gr.	n.	delight, grace	*Chaerophyllum*
chaet Gr.	n.	bristle	*Chaetopappa*
chamae Gr.	adv.	low, dwarf, on the ground	*chamaedrifolius*
cheil Gr.	n.	lip or margin	*Cheilanthes*
chen Gr.	n.	goose	*Chenopodium*
chers Gr.	n.	dry land	chersophilous
chil Gr.	n.	See *cheil.*	*Chilopsis*
chion Gr.	n.	snow	chionad
chlamyd Gr.	n.	a cloak	achlamydeous
chlo Gr.	n.	young shoots of grass, young herbage	*Echinochloa*
chlor Gr.	adj.	green, greenish-yellow	*Chloranthus*

Component	Type	Meaning	Example
chor Gr.	v.	scattering, dispersing	anemochore
chor Gr.	adj.	separate, distinct	choripetalous
chort Gr.	n.	fodder, forage	*Calochortus*
chromat Gr.	n.	color	chromatoplast
chron Gr.	n.	time	synchronous
chrys Gr.	adj.	golden, golden yellow	*chrysocomus*
cid L.	v.	to cut, to kill, to fall	septicidal, decidous
cili L.	n.	eyelashes, marginal hairs	ciliate
ciner L.	n.	ashes	*Cineraria*
circin Gr.	adj.	round, rolled up	circinate
ciss Gr.	n.	ivy	*Parthenocissus*
clad Gr.	n.	a young branch, a sprout, a shoot	cladophyll
clav L.	n.	a club	clavate
cleist Gr.	adj.	closed, shut	cleistogamous
co- L.	pref.	with, together	codominant
col L.	v.	to inhabit	*sylvicolus*
cole Gr.	n.	a sheath, scabbard	*Coleogyne*
com- L.	pref.	See *co-*.	*Comandra*
con- L.	pref.	See *co-*.	connate
confert L.	v.	dense, crowded	*confertifolius*
cord L.	n.	heart	cordiform
cori L.	n.	hide, leather	coriaceous
corn L.	n.	horn	corneous
cory Gr.	n.	a helmet, crest	*Corydalis*
cost L.	n.	a rib	intercostate
crass L.	adj.	thick, heavy	*crassifolius*
cren L.	n.	a notch	crenulate
creper L.	adj.	dark, gloomy, twilight	crepuscular
crotal Gr.	n.	rattle	*Crotalaria*
cruc L.	n.	a cross	*Cruciferae*
crypt Gr.	adj.	secret, hidden	*cryptandrus*
-culus L.	suf.	diminutive ending	*tuberculus*
cune L.	n.	a wedge	cuneate
curr L.	v.	run	decurrent
curt L.	adj.	shortened, clipped, broken	*curtipendula*
cuspid L.	n.	point	cuspidate
cyan Gr.	adj.	dark blue or black	*Cyanophyceae*
cycl Gr.	adj.	circular	*Cycloloma*
cyn Gr.	n.	a dog	*Cynoglossum*

D

Component	Type	Meaning	Example
dactyl Gr.	n.	a finger	*dactylifer*
dasy Gr.	adj.	hairy, shaggy	*Dasylirion*

Component	Type	Meaning	Example
de- L.	pref.	down, from, away	deflexed
dec, deca Gr.	adj.	ten	decapetalous
del Gr.	adj.	visible, manifest, clear	*Spirodela*
delta Gr.	n.	triangle, delta	*deltoides*
dendr Gr.	n.	a tree	dendrology
dens, dent L.	n.	a tooth	tridentate
derma Gr.	n.	skin	leucoderma
dermat Gr.	n.	See *derma.*	dermatogen
di- Gr.	pref.	double, two	diandrous
di- L.	pref.	asunder, apart, in various directions	dilate
dialy Gr.	v.	separate, disband	dialycarpic
diant Gr.	adj.	wetted	*Adiantum*
dich Gr.	adv.	twofold, apart	dichotomous
dif- L.	pref.	See *di-* L.	diffusion
digit L.	n.	a finger	*Digitaria*
diplo Gr.	adj.	twofold, double	*diplostegius*
dis- L.	pref.	See *di-* L.	disarticulate
dissaep L.	n.	partition, septum	dissepiment
diurn L.	adj.	of the day, by day, during the day	diurnal
divaric L.	v.	spread asunder, stretch apart	*divaricata*
divers L.	v.	turned in different or contrary directions	*diversiloba*
dodeca Gr.	adj.	twelve	dodecapetalous
drom Gr.	n.	a course, a running	dromotropism
dros Gr.	n.	dew	*Drosera*
drym Gr.	n.	a forest, oakwood, coppice	*drymaria*
dum L.	n.	a bramble, brier, thornbush	*dumosus*

E

Component	Type	Meaning	Example
e-, ex- L.	pref.	from, out of	emersed
-eae L.	pl. suf.	used to denote plant tribes	*Hordeae*
ec-, ex- Gr.	pref.	out of, from	*Ecballium*, *exanthem*
ec Gr.	n.	See *oec.*	ecology
echin Gr.	n.	a hedgehog (prickly)	*Echinochloa*
ect Gr.	adv.	outside, external	ectocarp
elae Gr.	n.	the olive tree	*Elaeagnus*
ele Gr.	n.	marsh	*Eleocharis*
-ell L.	suf.	diminutive ending	*Prunella*
elytr Gr.	n.	a cover, a sheath	*Hymenelytra*
end Gr.	adv.	within	endocarp
enne Gr.	adj.	nine	enneander
-ensis L.	suf.	belonging to a locality or region	*canadensis*

Component	Type	Meaning	Example
epi- Gr.	pref.	on, upon, over, above	epigynous
equ L.	n.	horse	*Equisetum*
er Gr.	n.	spring (of year)	*Erianthus*
erem Gr.	n.	desert, solitude	*Eremochloa*
eri Gr.	adv.	very	*Erigenia*
err L.	v.	to wander, stray	aberrant
erythro Gr.	adj.	red or reddish	*erythrospermus*
esc L.	n.	food	*esculentus*
-escens L.	part. ending	tending to, becoming	*viridescens*
-escent L.	part. ending	See *escens*.	canescent
-etum L.	suf.	place of a thing, or where they grow	arboretum
eu- Gr.	pref.	good, well, true	*Euthamia*
ex- Gr.	pref.	See *ec-*.	*exanthematicus*
ex- L.	pref.	See *e-*, L.	exserted
extra- L.	pref.	outside, additional, beyond	extravaginal

F

Component	Type	Meaning	Example
falc L.	n.	sickle, scythe	falcate
fasc L.	n.	a bundle, a bunch	fasciculate
-fer, -fera, -ferum L.	suf.	bearing, carrying	conifer
fil L.	n.	a thread	*filifolius*
filic L.	n.	a fern	filical
fimbr L.	n.	fringe	fimbriate
fistula L.	n.	a hollow reed, a pipe, a tube	*fistulosus*
flabell L.	n.	a small fan	flabellate
flect L.	v.	to bend	deflect
flocc L.	n.	tuft, lock	flocculate
flor L.	n.	a flower	floret
fluit L.	v.	to float, to swim	*fluitans*
fluvi L.	v.	flowing	*fluvialis*
foli L.	n.	leaf	foliaceous
foll L.	n.	a bag	follicle
form L.	n.	shape	*cordiformis*
frag L.	v.	breaking	*Saxifrage*
frond L.	n.	leafy branch, green bough, foliage	*frondosus*
fruct L.	n.	fruit	*rubifructus*
frument L.	n.	grain	frumentaceous
frutic L.	n.	a bush or shrub	fruticose

Component	Type	Meaning	Example
fulg L.	v.	to shine	*fulgidus*
fulv L.	adj.	yellow, reddish-yellow, tawny-golden	*fulvous*
funi L.	n.	a cord, rope	*funiculus*
furc L.	n.	fork	*furcatus*
fusc L.	adj.	brown, dusky, tawny	fuscous

G

Component	Type	Meaning	Example
gala Gr.	n.	milk	*Polygala*
galact Gr.	n.	milk, with milky fluid	*Galactodendron*
gam Gr.	n.	marriage, united	polygamous
ge Gr.	n.	earth, ground	geotropic
geiton Gr.	n.	a neighbor	geitonogamy
gemin L.	n.	twin, double	*geminatus*
gen Gr.	v.	birth	dermatogen
gen L.	n.	a knee	*geniculatus*
genesis Gr.	n.	origin, generation	parthenogenesis
ger L.	v.	to bear, to have	*Setigera*
geran Gr.	n.	crane (bird)	*Geranium*
gibb L.	adj.	protuberant	gibbous
gigant Gr.	adj.	giant, mighty	*giganteus*
glabr L.	adj.	smooth, hairless	glabrescent
gladi L.	n.	a sword	*gladiolus*
gland L.	n.	an acorn, a gland	*glandulosus*
glauc Gr	adj.	covered with a bloom	glaucous
glob L.	n.	ball	globose
glochin Gr.	n.	arrow-point	glochidium
glom L.	n.	a ball, a round body	*glomeratus*
gloss Gr.	n.	a tongue	*Ophioglossum*
glott Gr.	n.	See *gloss.*	*Glottidium*
glum L.	n.	a hull, husk	glume
glutin L.	n.	glue	*glutinosus*
glyc Gr.	adj.	sweet, pleasant	*Glyceria*
gon L.	n.	progeny, generation, begetting	archegonium
gon Gr.	n.	joint, knee, angle	*Polygonum*
gramin L.	n.	grass	gramineous
greg L.	n.	herd, flock	*aggregatus*
gymno Gr.	adj.	naked, lightly clad	gymnosperm
gyn Gr.	n.	female, ovary	gynandrous
gynec Gr.	n.	See *gyn*	gyneceum

Component	Type	Meaning	Example
H			
haben L.	n.	thong or strap	*Habenaria*
haer L.	v.	to stick, cling to	adhere
hama Gr.	adv.	all together, at the same time	*Hamamelis*
hapl Gr.	adj.	single, simple	haploid
hast L.	n.	a spear	hastate
hedy Gr.	adj.	sweet	*hedycarpus*
heli Gr.	n.	the sun	heliotropism
helv L.	adj.	honey-yellow	*helvolus*
hemer Gr.	adv.	for a day	hemeranthous
hemi- Gr.	pref.	half	hemicarp
hepat Gr.	n.	the liver	*Hepatica*
hesper Gr.	adj.	of an evening, west, western	*Hesperocallis*
heter Gr.	adj.	other, different	*heterolepis*
hiem L.	n.	winter	hiemal
hier Gr.	adj.	sacred	*Hierochloa*
hipp Gr.	n.	horse	*hippocastanus*
hirsut L.	adj.	rough, hairy	*hirsutissimus*
hisc L.	v.	to split open, yawn	dehiscence
hom Gr.	adj.	equal, same	homologous
homal Gr.	adj.	smooth, even, ordinary	homalium
humi L.	adv.	low	*humifusus*
hydr Gr.	n.	water	hydrophyte
hygr Gr.	adj.	moist, wet	hygroscopic
hymen Gr.	n.	a membrane	*Hymenocallis*
hyper- Gr.	pref.	over, above, beyond	hypertrophy
hypo- Gr.	pref.	under, below	hypogynous
hyps Gr.	n.	height	hypsometer
I			
-ill L.	suf.	diminutive ending	*pusillus*
ily Gr.	n.	mud, slime	*ilysanthes*
imbric L.	n.	hollow tile, trough	imbricate
-inae L.	pl. suf.	used to denote subtribes; of or pertaining to; like; characterized by	*Metastelmatinae*
-ineae L.	pl. suf.	used to denote plant suborders	*Malvineae*
-inus, a, um L.	suf.	of or pertaining to, characterized by	cauline
insul L.	n.	an island	*insularis*
inter- L.	pref.	between, among	internode
intra- L.	pref.	within	intravaginal
iso- Gr.	pref.	equal, same, similar	isotherm
-issimus L.	suf.	superlative degree ending	*angustissimus*

Component	Type	Meaning	Example
J			
jug L.	v.	join, marry, yoke	conjugation
L			
labi L.	n.	lip	*Labiatae*
lact L.	n.	milk	*lactifer*
lacu L.	n.	a lake	lacustrine
laen L.	n.	a cloak, a mantle	*Notholaena*
lago Gr.	n.	a hare	*Lagurus*
lampr Gr.	adj.	shining bright, beautiful	*Lamprotis*
lan L.	adj.	wool	*lanatus*
lance L.	adj.	spear, lance	lanceolate
lanugin L.	n.	down, soft hair	*lanuginosus*
lapp L.	n.	a burr	*lappulus*
lasi Gr.	adj.	shaggy with hair or wool	*lasiocarpus*
lepid Gr.	n.	See *lepis*.	*Lepidium*
lepis Gr.	n.	a scale	*heterolepis*
lept Gr.	adj.	slender, small, thin, weak	*Leptochloa*
leuc Gr.	adj.	white	*leucanthus*
levig L.	adj.	smooth, polished	*levigatus*
ligul L.	n.	a small tongue	liguliform
limn Gr.	n.	marsh, pond	*Limnanthes*
lin L.	n.	flax	*Linaria*
liri Gr.	n.	lily	*Liriodendron*
lith Gr.	n.	stone	*Lithospermum*
lob Gr.	n.	a lobe, a capsule, a pod	*Gonolobus*
loc L.	n.	a place, a case, a casket, a cell	locule
lochi Gr.	n.	childbirth, that which is born	*Aristolochia*
lochm Gr.	n.	thicket, coppice	lochmium
lomat Gr.	n.	fringe, border, hem	*Lomatium*
loph Gr.	n.	a crest, a tuft on crown of head	*Lophiola*
lophot Gr.	adj.	crested	*Lophotocarpus*
lun L.	n.	the moon	*Lunaria*
lute L.	adj.	clayish-yellow	lutescent
lyc Gr.	n.	wolf	*Lycoperdon*
lyg Gr.	n.	a plaint twig, a willow-like tree	*Lygodesma*
lyper Gr.	n.	pain, sorrow	*Lyperia*
lyr Gr.	n.	lyre	*lyratus*
lythr Gr.	adj.	defiled with blood or gore, dark red	*Lythrum*

Component	Type	Meaning	Example

M

Component	Type	Meaning	Example
macr Gr.	adj.	long, large	*macrospermus*
macul L.	n.	a spot, stain, mark	*maculatus*
magn L.	adj.	great, large	*magnificus*
malac Gr.	adj.	soft	*malacophyllum*
mamm L.	n.	breast, nipple, teat	*Mamillaria*
mar L.	n.	sea	*maritimus*
mas L.	n.	male	*masculinus*
mega Gr.	adj.	large	megaspore
mei Gr.	adj.	lesser, smaller	meiosis
mel Gr.	n.	honey	melitose
melan Gr.	adj.	black	*melanocarpus*
men Gr.	n.	moon	*Menispermum*
mer Gr.	adj.	a part	pentamerous
meso- Gr.	pref.	in the middle, midway between	mesophyte
meta- Gr.	pref.	along with, between, after	metachlamydeous
micro Gr.	adj.	little, small	*micranthus*
migr L.	adj.	to move, depart	migration
mill L.	adj.	a thousand, very many	*millefolius*
mim L.	n.	a mimic, mime	*mimulus*
minth Gr.	n.	mint	*Acanthomintha*
mir L.	adj.	wonderful, extraordinary, remarkable	*Mirabilis*
mitr Gr.	n.	a head band, a turban	*mitriostigmus*
moll L.	adj.	soft	*Mollugo*
monil L.	adj.	a necklace, string of beads	moniliform
mono- Gr.	pref.	one, single, alone	*monolepis*
morph Gr.	n.	shape or form	dimorphic
mucr L.	n.	an abrupt point or tip	mucronate
mult L.	adj.	much, many	multilocular
mur L.	n.	mouse	*murinus*
myos Gr.	n.	of a mouse	*Myosurus*
myr Gr.	n.	ointment, balsamic juice, sweet oil	*Myristica*

N

Component	Type	Meaning	Example
nast Gr.	adj.	close pressed	nyctinastic
ne- Gr.	pref.	not, without, free from	*nepenthes*
nem, nemat Gr.	n.	thread, yarn	*Nemastylis*
neo- Gr.	pref.	new, young	neophyte
nephr Gr.	n.	kidney	*nephrolepis*

Component	Type	Meaning	Example
neur Gr.	n.	nerve, vein, string	platyneuron
nict L.	n.	wink	*nictitans*
nid L.	n.	a nest	*Nidulacium*
nigr L.	adj.	black, dark	nigrescent
noct L.	n.	night	nocturnal
nomen L.	n.	name	nomenclature
not Gr.	adj.	the back, rear	nototribal
noth Gr.	adj.	spurious, bastard	*Notholaena*
nud L.	adj.	naked, uncovered	*nudicaulis*
nyct Gr.	n.	night	nyctotropic
nymph Gr.	n.	a minor goddess of waters	*Nymphaea*

O

Component	Type	Meaning	Example
ob- L.	pref.	inversely, against	oblanceolate
ochr Gr.	adj.	pale yellow	ochraceous
octo- L.	pref.	eight	*octoflorus*
-odes Gr.	suf.	See *oides.*	*trichodes*
oec, ec Gr.	n.	house	dioecious
oic Gr.	n.	See *oec.*	*dioicus*
-oid Gr.	suf.	See *oides.*	deltoid
-oideae L.	pl. suf.	ending for plant subfamilies	*Panicoideae*
-oides Gr.	suf.	form of, type of, similar to	*Seirpoides*
olens L.	v.	smelling, sweet-smelling	*graveolens*
-olus, a, um L.	suf.	diminutive ending	*gladiolus*
ont Gr.	n.	a single being, an individual	ontogeny
onym Gr.	n.	name	synonym
op Gr.	n.	milky fluid, plant juice	*opopanax*
opor Gr.	n.	the end of summer, fruiting time	*oporanthus*
opsis Gr.	n.	aspect, view, appearance	*Oryzopsis*
or Gr.	n.	mountain	orographic
orb L.	n.	ring, circle, disc	*orbicularis*
ornith Gr.	n.	a bird	ornithogamous
orth Gr.	adj.	straight	orthotropous
orthr Gr.	n.	dawn, about daybreak	*Orthrosanthus*
osm Gr.	n.	smell, scent	*Hedyosma*
-osus L.	suf.	full of, abounding in	*filamentosus*
ur Gr.	n.	tail	*Myosurus*
ox, oxy Gr.	n.	sharp, keen, quick, acid	*oxylepis*, *oxalis*

Component	Type	Meaning	Example
P			
pachy Gr.	adj.	thick	pachycladous
palud L.	n.	marsh, swamp	*paludosus*
pand L.	v.	spread, expand, unfold	repand
papill L.	n.	a nipple, a teat	papillose
papyr Gr.	n.	rush, papyrus, hence paper	*papyrifer*
para Gr.	prep.	beside, near, parallel	paracarpous
pariet L.	n.	wall	*Prietaria*
parm L.	n.	a small round shield, a target	*Parmelia*
parthen Gr.	adj.	virgin, pure	*Parthenocissus*
parv L.	adj.	little, small	*parvifolius*
pasch Gr.	n.	passion	*Paschoanthus*
passi L.	n.	passion, suffering	*Passiflora*
pauc L.	adj.	few	*pauciflorus*
pectin L.	n.	comb	*pectinatus*
ped Gr.	n.	a plain, level country	*Merismopedia*
ped L.	n.	foot, base	*pedatus*
pelt Gr.	n.	a small shield	peltate
pend L.	v.	to hang down	*curtipendulus*
penn L.	n.	feather, pen, wing	*Pennisetum*
pent Gr.	adj.	five	pentandrous
per- L.	pref.	through, very	perennial, *peramoenus*
peri- Gr.	pref.	around	pericarp
persic Gr.	n.	peach	*Lycopersicon*
pet L.	v.	go toward, seek	centripetal
phall Gr.	n.	the penis, testes	*Amorphophallus*
phaner Gr.	adj.	visible, manifest	phanerogam
phil Gr.	adj.	loving, fond of	entomophilous
phob Gr.	n.	fear, flight	calciphobia
phor Gr.	n.	a thief	*Phoradendron*
phor Gr.	v.	bearing, carrying	gynophore
phyl Gr.	n.	race, stock	phylogeny
phyll Gr.	n.	leaf	macrophyllous
phyt Gr.	n.	plant	phytologist
pil L.	n.	hair	piliferous
pinn L.	n.	feather	pinnatifid
platy Gr.	adj.	broad, wide, flat	*platycarpus*
plei Gr.	adj.	more, many	pleiopetalous
pleur Gr.	n.	side or rib	pleurotribal
ploc Gr.	n.	lock of hair, curl, wreath	*Symplocus*
plur L.	adj.	several, many	plurilocular
pod Gr.	n.	foot	*Podophyllum*
pogon Gr.	n.	beard	*Andropogon*

Component	Type	Meaning	Example
poly Gr.	adj.	many, much	polypetalous
por Gr.	n.	opening, passage	poricidal
potam Gr.	n.	river	*Potomogeton*
prae- L.	pref.	before, in front, early	praecocious
prim L.	adj.	first	*primulus*
pro- Gr.	pref.	before, in front of, forward	proanthesis
proto- Gr.	pref.	first, primary	protandrous
psamm Gr.	n.	sand	psammophilous
pseud Gr.	adj.	false, deceptive	*pseudoacacius*
psil Gr.	adj.	naked, smooth	*Psilostrophe*
psor Gr.	n.	itch, mange, skin disease	*Psoralea*
psychr Gr.	n.	cold	psychrometer
pter Gr.	n.	wing	pterocarpous
pterid Gr.	n.	male fern, wing	*Pteridophyta*
pub L.	n.	the hair of puberty	pubescent
pulchr L.	adj.	beautiful	*pulcherrimus*
pulv L.	n.	dust	*pulverulentus*
punct L.	adj.	point or dot	*punctatus*
pycn Gr.	adj.	dense, close, compact	*pycnostachya*
pyr Gr.	n.	fire	*Pyracantha*
pyr, pir L.	n.	pear	pyriform
pyr Gr.	n.	wheat	*Agropyron*
pyrrh Gr.	adj.	flame-colored, reddish	*Pyrrhopappus*

Q

Component	Type	Meaning	Example
quadri L.	adj.	four	*quadrifolius*
quinque L.	adj.	five	*quinquefolius*

R

Component	Type	Meaning	Example
rach Gr.	n.	a spine, backbone, axis, ridge	rachilla
ram L.	n.	a branch	*ramulosus*
ran L.	n.	a frog	*Ranunculus*
re- L.	pref.	back, again	reflexed
ren L.	n.	kidney	reniform
ret L.	n.	a net	reticulate
retr L.	adj.	backward	*retroflexus*
rhach Gr.	n.	See *rach.*	rhachilla
rhiz Gr.	n.	root	rhizoid
rhod Gr.	n.	a rose	*Rhododendron*

Component	Type	Meaning	Example
rhynch Gr.	n.	a beak, snout	rhynchosporous
rigid L.	adj.	stiff, inflexible	rigidulous
rip L.	n.	the bank of a stream	riparian
riv L.	n.	stream	*rivularis*
rostr L.	n.	a snout, beak, or bill	rostellum
rot L.	n.	a wheel	rotiform
rotund L.	adj.	round	*rotundifolius*
rub, rubr L.	adj.	red, ruddy	rubescent
ruf L.	adj.	red	*rufidulus*
rug L.	n.	wrinkle, crease	*rugosus*
rup L.	n.	rock	*rupestris*

S

Component	Type	Meaning	Example
sacc L.	n.	a sac, a bag	saccate
sacchr Gr.	n.	sugar	*saccharoides*
sagitt L.	n.	an arrow	*sagittifolius*
salic L.	n.	willow	*salicifolius*
sangu L.	n.	blood	*Sanguinarius*
saur Gr.	n.	lizard	*Saururus*
sax L.	n.	stone, rock	*Saxifraga*
scabr L.	adj.	rough	scabrous
schiz Gr.	v.	to split	schizocarp
scop L.	n.	twig, shoot, broom	*scoparius*
scord Gr.	n.	plant with garlic odor	*Nothoscordum*
seb L.	n.	suet, grease, tallow	sebaceous
sect L.	v.	cut	bisect
semper L.	adv.	forever, always	*sempervirens*
senec L.	adv.	an old man, old	*Senecio*
sep L.	n.	hedge, partition, fence	*Sepium*
sept L.	n.	wall, enclosure	septifragal
ser L.	adj.	late, slow	*serotinus*
serr L.	n.	saw	serrate
set L.	n.	a bristle, a hair	*Setaria*
silv L.	n.	a wood, forest	*silvaticus*
siphon Gr.	n.	a tube	*Stenosiphon*
sobol L.	n.	sprout, offshoot	soboliferous
sor Gr.	n.	a heap	sorus
sperm Gr.	n.	seed	gymnosperm
sphaer Gr.	n.	sphere, ball	*sphaerocarpus*
spic L.	n.	a point, spike	spicate
spin L.	n.	a thorn, spine	spinulous
spir Gr.	n.	a coil, spiral, twist	spiricle

Component	Type	Meaning	Example
spor Gr.	n.	seed, spore	sporophyll
squam L.	n.	a scale	*squamosus*
steg Gr.	n.	cover, roof	*Physostegia*
stell L.	n.	a star	*Stellaria*
stemm Gr.	n.	a crown, wreath, garland	*Agrostemma*
sten Gr.	adj.	narrow	*stenandrius*
steph Gr.	n.	wreath, to surround	*Androstephium*
stich Gr.	n.	a line or row of things	*Polystichium*
stoma, stomat Gr.	n.	mouth	stomatic
strept Gr.	adj.	bent, pliant, twisted	*streptopus*
stri L.	n.	a furrow, channel	*striatus*
strob Gr.	n.	a top, a whirling around	*strobilus*
stroph Gr.	v.	turning, twisting	*strophostylus*
styl Gr.	n.	style, stake, pillar	*Stylosanthes*
suav L.	adj.	sweet, pleasant	suaveolent
sub- L.	pref.	under, below, close to	subsessile
subul L.	n.	an awl	subulate
sulc L.	n.	a furrow, groove	sulcate
super- L.	pref.	over, above, on top	superaxillary
supin L.	adj.	lying on the back	resupinate
supra- L.	pref.	above, beyond, before	supranodal
sylv L.	n.	a wood, forest	*sylvaticus*
sym- Gr.	pref.	with, together	symbiosis
syn- Gr.	pref.	See *sym-*.	synema

T

Component	Type	Meaning	Example
tax Gr.	n.	order, arrangement	phyllotaxy
tel Gr.	n.	end, purpose, completion	teleology
tele Gr.	adj.	far, far off, at a distance	*Telopea*
tenu L.	adj.	thin, narrow, slender	*tenuifolius*
tephr Gr.	n.	ashes	*Tephrosa*
tern- L.	pref.	in threes, three at a time	ternate
terr L.	n.	earth	*terrestris*
tetr Gr.	adj.	four	tetrandrous
tham Gr.	adj.	crowded, close	*Euthamia*
thamn Gr.	n.	copse, thicket, bush	*Chrysothamnus*
thel Gr.	n.	nipple, teat	*thelephorus*
therm Gr.	n.	heat	thermotaxis
thes Gr.	n.	a putting or setting in order	photosynthesis
thet Gr.	n.	See *thes*.	synthetic

Component	Type	Meaning	Example
thigm Gr.	n.	touch	thigmotropism
tinct L.	v.	dyed	*tinctorius*
tom Gr.	v.	cutting	dichotomy
tor L.	n.	swelling, protuberance, bed, couch	torus
tort L.	adj.	twist, bend	*contorta*
toxic Gr.	n.	poison	*Toxicodendron*
trachy Gr.	adj.	rough, shaggy	*trachycarpus*
trag Gr.	n.	a male goat	*Tragopogon*
tri- L., Gr.	pref.	three, in three parts	trifoliate
trich- Gr.	pref.	in three parts	trichotomous
trich Gr.	n.	hair	*Trichomanes*
trop Gr.	v	turning, changing, bending	heliotropic
turg L.	v.	swell	turgescent
typo L.	n.	model, pattern, figure	ecotype

U

Component	Type	Meaning	Example
umbell L.	n.	a little shade (parasol)	umbelliferous
un L.	adj.	one, single	*uniflorus*
unc L.	n.	a hook, barb	*uncinatus*
urce L.	n.	pitcher	urceolate

V

Component	Type	Meaning	Example
vacc L.	n.	cow	*Vaccinium*
vagin L.	n.	a sheath	*vaginaeflorus*
vall L.	n.	valley, groove	vallicula
ver L.	n.	spring (season)	vernal
verruc L.	n.	a wart	*verrucosus*
vers L.	adj.	turning easily, versatile	*versutus*
vertic L.	n.	a whorl	*verticillata*
vesic L.	n.	a blister, bladder	*vesicarius*
vesper L.	n.	evening, west	*vespertinus*
vest L.	v.	to cover, clothe	*vestitus*
vill L.	n.	long shaggy hair	villous
vir L.	adj.	green	*virescens*
virg L.	n.	a thin green twig or wand	*virgatus*
volv L.	v.	to roll, turn, twist	*convolvulus*
vulp L.	n.	fox	*vulpinus*

Component	Type	Meaning	Example
X			
xanth Gr.	adj.	yellow	*Xanthium*
xen Gr.	n.	guest, stranger	xenogamy
xer Gr.	adj.	dry	xerophytic
Z			
zo Gr.	n.	alive, living, animal	zoöspore
zyg Gr.	n.	yoke	zygomorphic

Appendix

Floral Evolution

In the evolution of the higher plants, the greatest number of changes has come about in the reproductive organs. This makes the flower the most important part of the plant from the viewpoint of the taxonomist. Most of these changes can be classified under the following headings:

1. *Change in number plan and reduction.* Flowers low in the scale of evolution usually have many parts (an indefinite number), as for example: The Ranunculi have many stamens and many pistils, whereas the legumes usually have ten stamens and one pistil.
2. *Change in union.* Flowers low in the scale of evolution have their parts separate (not grown together), while the more advanced ones have their parts coalesced (grown together), as for example: The Ranunculi have separate petals, separate stamens, and separate pistils, whereas in the Ipomoëae, the petals are grown together, stamens are attached to the corolla, and the pistil is composed of coalesced carpels.
3. *Change in shape.* The parts of the more primitive flowers are usually simple and all of the members of a series are similar, whereas the more recent flowers tend to have complex parts and members of a series may vary in shape, as for example, the actinomorphic flowers of *Calytonia* and the zygomorphic flowers of *Scrophularia.*

4. *Change in elevation.* In the more primitive flowers, the sepals, petals, and stamens are attached below the ovary (hypogynous), while in the higher plants these are attached either around (perigynous) or on top of the ovary (epigynous).
5. *Change in arrangement.* The spiral arrangement of floral parts is considered more primitive than the cyclical (whorled) arrangement.

In summarizing these changes, it may be said that flowers low in the scale of evolution have parts *numerous* (number indefinite), *simple, separate, spirally arranged,* and *hypogynous;* while those high in the scale have parts *few* (number definite), *complex, grown together, cyclically arranged,* and *epigynous.*

Laws, Theories, and Hypotheses

These laws, theories, and hypotheses are more or less related to taxonomy, evolution, and speciation.

Bergman's Law. The maximum size of a species is found in the optimal region of its range.

Chapman's Law of Biotic Potential. The biotic potential of a species is a quantitative expression of the dynamic power of the species which is pitted against the resistance of the environment in which it lives in its struggle for existence.

Chapman, Royal N. *Animal Ecology,* p. 183. McGraw-Hill, 1931.

Darwin's Theory of Evolution. Natural selection through survival of the fittest. The species best suited to its environment has the best chance for survival.

deVries' Mutation Theory. New species and forms arise as unusual abrupt deviations from their parents, and breed true for that deviation.

Dollo's Law of Irreversibility. An organism never reverts exactly to its orignal form, even if returned to its original environment.

Fernald's Nunatak Hypotheses. Arctic-alpine plants persisted on ice-free lands in the arctic and upon small unglaciated areas (nunataks) well within or near the margins of the great ice fields.

Raup, Hugh M. Botanical problems in boreal America. I. Bot. Rev. 7., p. 183. 1941.

Haeckel's Biogenetic Law. (*Recapitulation Theory*). Ontogeny recapitulates phylogeny. That every organism in its individual life-history repeats the various stages through which its ancestors have passed in the course of evolution.

Hopkins' Bioclimatic Law. Other things being equal, the variation in the time of occurrence of a given periodic event in the life activity in temperate North America is at the general average rate of four days for each degree of latitude, five degrees of longitude, and 400 feet altitude; later northward, eastward, and upward in spring and early summer; and the reverse in late summer and autumn.

Hopkins, A. D. Periodical events and natural law as guides to agricultural research and practice. U.S.D.A. Monthly Weather Rev., Supp. No. 9, 42 pp. 1918.

Jordan's Law of Geminate Species. Twin species, each representing the other on opposite sides of some form of barrier.

Jordan, David Starr. Amer. Naturalist, Vol. 42, pp. 73-80. 1908.

Knight-Darwin Law. No organic being fertilizes itself for an eternity of generations. Nature abhors perpetual self-fertilization.

LeChatelier's Theorem. Every change in the direction of an intensification of the environmental conditions influencing a body or a system of bodies, augments the resistance of the latter to a further increase of this influence.

Maximov, N. R. *The Plant in Relation to Water,* p. 326. Macmillan Company, 1929.

Liebig's Law of the Minimum. When a multiplicity of factors is present and only one is near the limits of toleration, this factor will be the controlling one.

Lotsy's Theory of Hybridization. Species arise by crossing, perpetuate themselves by heredity and are gradually exterminated by the struggle for life, those last exterminated being the selected ones.

Lotsy, J. P. *Evolution by Means of Hybridization*, p. 157. Martinus Nijhoff, The Hague, 1916.

McAtee's Survival of the Ordinary. That reproduction of species, on the whole, is carried on by ordinary individuals. There are more ordinary individuals than there are other; therefore, they have more chance to survive.

McAtee, W. L. Survival of the ordinary. Quarterly Rev. of Biol., Vol. 12, pp. 47–64. 1937.

The Doctrine of Signatures. An old belief that medicinal plants were marked in some way to indicate their value to man; the problem was to discover the sign and interpret it properly. In application, any part of a plant that resembled a part of the human body in form, texture, or color was usually considered to have curative qualities for all of the ills of that organ.

The Law of Compensating Factors. A factor that is weak or functionless in a complex organism is often supplemented or replaced by the modification of other factors when it is advantageous to the species.

Vavilov's Distribution of Variabilities. The greatest concentrations of variability in genera or species are found in their areas of origin, where differentiation has been taking place for the longest time. At the periphery of the area occupied by a plant and in places of natural isolation one is most likely to find unusual forms of that plant, due to inbreeding or mutation.

Vavilov's Law of Homologous Series. "Species and genera that are genetically closely related are characterized by similar series of heritable variations with such regularity that, knowing the series of forms within the limits of one species, we can predict the occurrence of parallel forms in other species and genera. The more closely related the species in the general system, the more resemblance will there be in the series of variations. Whole families of plants in general are characterized by definite cycles of variability occurring through all genera and species making up the family."

English translation in *Selected Work of N. I. Vavilov* (K. Starr Chester, Chronica Botanica Publication, 1949).

Wagner's Isolation Theory (As stated by Jordan and Kellog). Given any species, in any region, the nearest related species is not to be found in the same region nor in a remote region, but in a neighboring district separated from the first by a barrier of some sort or at least a belt of country, the breadth of which gives the effect of a barrier.

Jordan, David Starr, and Vernon L. Kellog. *Evolution and Animal Life*, p. 120. D. Appleton Co., 1908.

Willis's Age and Area Hypothesis. In general, the area occupied by a species depends upon its age, or, conversely, the age of a species is proportional to its geographical area.

Willis, John C. The age and area hypothesis. Science, Vol. 47, pp. 626–28. June, 1918.

Zalenski's Law. The anatomical structure of the individual leaves of a shoot is a function of their distance from the root system.

Arber, Agnes. *The Gramineae*, p. 305. Macmillan Company, 1934.

Phyllotaxy

(Leaf Arrangement)

The disposition of leaves on stems usually follows some definite arrangement. Plants with only one leaf at a node are said to be *alternate*. Those with two or more are termed *whorled* or *verticillate*. The simplest whorl consisting of two leaves is said to be *opposite*.

Alternate leaves are usually arranged spirally. This spiral may turn clockwise on some species and counterclockwise on others. Sometimes the spiral may be interrupted, but it is sufficiently constant to convince one that there must be some rule involved.

The Italian mathematician, Leonardo of Pisa, surnamed Fibonacci, formulated a series of numbers known as the Fibonacci summation series. This series, 0, 1, 1, 2, 3, 5, 8, 13, 21, etc., is formed by using the sum of the last two consecutive numbers as the next higher number of the series. M. A. Brown, with other prominent European botanists who were studying the arrangement of leaves on stems, found that by combining the Fibonacci summation series and the principle of the "Spirals of Archimedes," they had a fractional series that would apply to most of the alternate leaf arrangements. This series is known as the Fibonacci fractional series. The fractions are: 1/2, 1/3, 2/5, 3/8, 5/13, 8/21, 13/34, 21/55, etc., and are formed by taking the series twice, placing one above the other and two places to the right. This will make the first "1" in the top series appear above the "2" in the bottom series, the second "1" above the "3," etc. The sum of any two consecutive numerators equals the next higher one in the scale, and the same is true for the denominators.

In application of the fraction series, the numerator indicates the number of complete turns the spiral makes and the denominator the number of consecutive leaves touched by the spiral. For instance, corn has leaves in two ranks. If a spiral is started at one leaf and continued up the stem touching all leaves in order of their elevation until one is reached that is directly above the first, the spiral will have made one complete turn and included two leaves (not counting the top one): thus the phyllotaxy of corn would be represented by the fraction ½. In the case of poplars, the end of the spiral at one complete turn does not stop on a leaf, so it is necessary to continue the spiral until it does stop on a leaf directly above the one used as a starting point. This is accomplished at the completion of the second complete turn. In making the two turns, five leaves are passed, thus making it a ⅖ arrangement. The following are some examples of the above series.

1/2-*Ulmus, Betula, Gramineae*
1/3-*Alnus*, sedges
2/5-*Quercus, Populus*
3/8-Holly, aconite, *Ailanthus*
5/13-*Pinus strobus* cone, rosettes of houseleek
8/21-Uncommon
13/34-Some pine cones
21/55-Some pine cones

Some less common leaf arrangements belong to other series, such as 1/4, 1/5, 2/9, 3/14, etc., but they follow the same law of the summation of the last two figures being the next higher one in the series.

Bibliography

BAILEY, L. H., 1933. How plants get their names. Macmillan Co., New York.

———, 1949. Manual of cultivated plants. Revised Ed. Macmillan Co., New York.

CAIN, STANLEY R., 1944. Foundations of plant geography. Harper and Bros., New York.

CARPENTER, J. RICHARD, 1938. An ecological glossary. Univ. Oklahoma Press. Norman, Okla.

CONINCK, A. M. C. JOHGKINDT, 1926. Dictionnaire Latin-Grec-Français-Allemand-Hollandais des principaux termes. Repr. 2nd. ed. Stechert, New York.

DAVIES, P. A., 1939. Leaf position in *Ailanthus altissima* in relation to the Fibonacci series. Amer. Jour. Bot., Vol. 26, pp. 67–74.

DAYTON, W. A., 1950. Glossary of botanical terms used in range research. U.S.D.A. Misc. Publ. No. 110. (First ed. July, 1931. Revised June, 1950.)

THE ECOLOGICAL SOCIETY OF AMERICA, 1938. Report of Committee on Nomenclature. 24th Annual Meeting, Richmond, Va.

FERNALD, M. L., 1950. Gray's manual of botany. 8th. ed. Amer. Book Co., New York.

GRAY, A., 1897. Gray's botanical textbook. Vol. I, Structural botany. Amer. Book Co., Chicago.

HARLOW, WM., AND E. S. HARRAR, 1941. Textbook of dendrology. McGraw-Hill, New York.

HENDERSON, I. F., AND W. D. HENDERSON, 1949. Dictionary of scientific terms. Van Nostrand, New York. (First ed. 1929. 4th ed. revised by John H. Kenneth, 1949.)

HITCHCOCK, A. S., 1950. Manual of the grasses of the United States. U.S.D.A. Misc. Publ. 200. Govt. Printing Office, Washington, D. C. (First ed. 1935. 2nd ed. revised by Agnes Chase, 1950.)

JACKSON, B. D., 1949. A glossary of botanic terms. J. B. Lippincott Co., Philadelphia. (First ed. 1928. Re-issued 1949.)

JEAGER, EDMUND C., 1944. A source-book of biological names and terms. Chas. C. Thomas, Pub., Springfield, Ill. (Second ed. 1950.)

KNIGHT, R. L., 1948. Dictionary of genetics. Chronica Botanica, Waltham, Mass.

LAWRENCE, GEORGE H. M., 1951. Taxonomy of vascular plants. Macmillan Co., New York.

LEAVITT, R. G., 1901. Outlines of botany. Amer. Book Co., New York.

MELANDER, A. L., 1940. Source book of biological terms. Biol. Dept., College of the City of New York.

SARGENT, C. S., 1926. Manual of the trees of North America. Houghton Mifflin Co., New York.

SWINGLE, DEANE B., 1946. A textbook of systematic botany. 3rd ed. McGraw-Hill, New York.

WOODS, R. S., 1944. The naturalist's lexicon. Abbey Garden Press, Pasadena, Calif.

———, 1947. Addenda to the naturalist's lexicon. Abbey Garden Press, Pasadena, Calif.

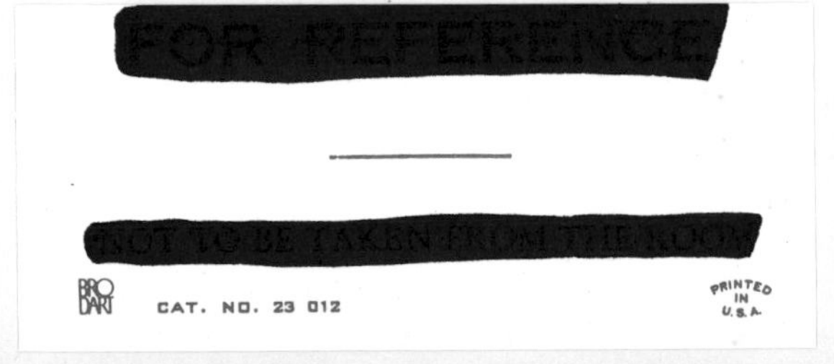
FOR REFERENCE
NOT TO BE TAKEN FROM THE ROOM
BRO DART
CAT. NO. 23 012
PRINTED IN U.S.A.